BIBLICAL WORDS FOR TIME

STUDIES IN BIBLICAL THEOLOGY

# BIBLICAL WORDS
# FOR TIME

JAMES BARR

SCM PRESS LTD
56 BLOOMSBURY STREET
LONDON

FIRST PUBLISHED 1962
© SCM PRESS LTD 1962
PRINTED IN GREAT BRITAIN BY
ROBERT CUNNINGHAM AND SONS LTD
LONGBANK WORKS, ALVA

# CONTENTS

# PREFACE

HERE I need only express my thanks to those who have helped me with discussion and advice. I would mention firstly the Rev. John McIntyre, Professor of Divinity in this University, and I must express the satisfaction with which I have found that certain positions which I reached after working on detailed linguistic points showed important agreements with positions which he had reached by the methods of the philosophy of religion. Among other colleagues I must mention with gratitude Mr Michael C. Stokes, Lecturer in Greek, whose opinion about Greek usage has been of the utmost value to me, and the Rev. R. A. S. Barbour, Senior Lecturer in New Testament. Outside of this University, I have been substantially helped by the advice of the Rev. C. F. D. Moule, Lady Margaret's Professor of Divinity at Cambridge, of the Rev. P. R. Ackroyd, Professor of Old Testament Studies at London, and of Mr Cyril Aldred of the Royal Scottish Museum.

*The University*                                                     J. B.
*Edinburgh*
*Easter* 1961

*This book is dedicated to the memory of*
OLIVER SHAW RANKIN
*who was the author's teacher in Old Testament*
*and to whose Chair he later succeeded*

זכרונו לברכה

quid autem familiarius et notius in lo-
quendo conmemoramus quam tempus? et
intellegimus utique, cum id loquimur, intellegimus etiam, cum alio loquente id
audimus. quid est ergo tempus? si nemo
ex me quaerat, scio; si quaerenti explicare
velim, nescio.

St Augustine, *Conf.* xi.14

# I

## INTRODUCTORY

It would be widely agreed that the last thirty years have seen a great revival of biblical theology. In general, this means a renewed attention to the theological purpose of the biblical writers, and a less exclusive concentration on the questions of historical criticism which had so occupied the preceding period. It involves no disparagement of historical criticism, but an awareness that something more is required, and in particular something which will bring out the coherence or unity of the biblical material, in contrast with the rather analytical methods of historical criticism. It suggests, moreover, that a certain lack of appreciation of the theological consciousness of the biblical writers, and in particular of the unity of that consciousness, has sometimes caused historical criticism to pose false questions, or to offer poor answers to real questions.

While it is true, therefore, that the revival of biblical theology may be in general understood simply as a renewed attention to the theological significance of the Bible, in fact it has involved certain methods and emphases, of which the first is the interest in bringing out the unity and coherence of the Bible.

The second such interest is that in bringing out the distinctiveness of the Bible. This is indeed in part a corollary of the first, for it is by setting the Bible in contrast with non-biblical thought that its own inner coherence is emphasized. The coherence and the distinctiveness of biblical thought are thus correlative and interdependent in the method. But the establishment of the distinctiveness of biblical thought is not only a necessary part of the method; it is also a principal purpose of the method.

Special attention should be given to one aspect which is commonly strongly emphasized. One of the chief ways in which the distinctiveness of the Bible is often brought out is through the contrast between Hebrew and Greek thought. The general lines of this contrast, as it is frequently presented, have been set out by the writer elsewhere,[1] and need hardly be repeated here, since they are in any case very familiar. The New Testament

[1] *The Semantics of Biblical Language*, pp. 8-14.

belongs, it is believed, entirely or almost entirely to the solid block of Hebrew thought, and thus the whole Bible stands in monolithic solidarity against Greek thought.

While there have, of course, been exceptions and differences of emphasis, it would hardly be questioned, the writer submits, that the features stated above have been extremely powerful, or dominant, in the modern revival of biblical theology, and typical of it. While it could indeed be argued that the term 'biblical theology' could be used simply of an appreciation of the theological significance of the Bible, it will surely not be denied that the term is normally used in the English-speaking theological world to indicate a new and distinctive approach, which had its modern beginning in the 1930's, and which has among its chief characteristics the features mentioned above. Moreover, if one were to ask an enthusiastic supporter of modern biblical theology for examples of the methods and results typical of the movement at its best, I think it undeniable that material like the works of Marsh, Robinson and Cullmann, which are to be examined in this study, would very probably be included. The general handling of the problems, which these books display, bears all the marks which characterize the modern theological movement in biblical interpretation and which mark most clearly its contrast with the earlier 'historical-critical' approach.

The direct purpose of this present book, however, is not to discuss or to criticize the main interests and concerns of biblical theology, such as the unity and distinctiveness of biblical thought and its basic differences from Greek thought. Its purpose is a much more limited one: to examine one particular procedure, and only one, which has in fact been used within the general context of this modern theological approach. This procedure is the building of a structure from the lexical stock of the biblical languages, and the assumption that the shape of this structure reflects or sets forth the outlines of biblical thinking about a subject. The subject in this case is time. I hope to show that this procedure is, in this case at any rate, an entirely faulty one. While the main interests and concerns of modern biblical theology are not a direct subject of discussion in this book, it is the opinion of the writer that these basic positions have not uncommonly been stated or used in ways which have led to defective methods in handling linguistic evidence. The more general statement, exemplification and criticism of these defective methods has been provided by the writer in his *The Semantics of Biblical Language*. The present work is a detailed study of one such procedure; one which has indeed been briefly mentioned in that former work, but which requires much more detailed investigation, and which, for reasons which will emerge later, has had particularly serious effects in the study of the subject of time in the Bible. The criticisms of this present work, then, do not directly affect the approach of biblical theology except in so far as the particular procedure criticized has been used.

Indirectly, however, it is true that the criticisms here made raise further questions about the interpretation of the Bible, and about its relation to philosophical and dogmatic theology. Firstly, the procedure criticized here is criticized precisely because the writer believes it to depend on, and to follow from, assumptions in the treatment of biblical language which have become in recent theology not only increasingly widespread but also increasingly regulative for biblical interpretation. It is not a procedure which *happens* to be faulty, but one the faultiness of which is representative and typical of certain tendencies in biblical interpretation. It must be remembered that it is entirely in the interests of modern theological interpretation that its methods of handling linguistic evidence should be examined with all strictness. Certain habits in the treatment of linguistic data (and I would mention in particular the great influence of Kittel's dictionary) have in fact contributed a very great deal to the picture now common of the distinctiveness and the coherence of the Bible. If these habits are faulty in any material way, and if our prime interest is in the right interpretation of the Bible, then it is most desirable that we should find ways of re-stating the message of the Bible and its character in a manner which does not depend on faulty procedures, and which therefore will reflect more truly the intention of the biblical writers.

Secondly, our discussion may indirectly raise questions about the scope and extent of what can be validly determined by appeal to the Bible. It is possible that in certain cases dubious linguistic methods have been adopted unconsciously because they were, or seemed to be, the only way in which the biblical material could appear to answer a given problem, or at any rate to answer it when posed in a particular way. If this should prove to be the case, it might mean that it was mistaken to include this problem within the scope of biblical exegesis at all, and that the handling of it should lie not directly within the biblical area but rather within a philosophical theology, which, while recognizing biblical authority, would not claim to be biblical exegesis. We should do well to remember that we do no honour to the Bible by wringing from its words answers to questions which were not in fact in the minds of those who used these words, and that by so doing we may only obscure the undivided attention with which the biblical writers concentrated upon their own themes.

Such, then, is the general theological setting of our investigation. Although the detailed study lies only in one small corner of the field of biblical interpretation, the issues ultimately raised include quite basic matters of method in biblical interpretation, along with the integration of biblical study with dogmatic and philosophical theology, and therefore along with the general organization of theological research. This general setting, having been thus mentioned, may now be left in the background, and returned to after we have done our detailed work on the interpretative procedure to be studied.

We now turn to more specifically linguistic questions. Although it is perfectly possible to make assertions about the coherence and distinctiveness of the Bible without any reference to its language in particular, that is, to characteristics of the Hebrew of the Old Testament and of the Greek of the New, it has in fact been common for such assertions to be linked with assertions about these language systems. Thus it is widely believed, to begin with, that the layout of the lexical stock of biblical Hebrew constitutes a reflection or adumbration of the theological thought of the Israelites, and thus ultimately of the actual theological realities acknowledged in that thought. On the one hand the gathering by an appropriate lexical procedure of the more important elements in the lexical stock can help to show forth the unity and coherence of the Old Testament. Although the documents are of different ages, sources, and interests, the same stock of terms is found in them all (apart of course from words peculiar to particular authors). On the other hand the elements thus assembled, and the mental realities or concepts involved in them, can be held, and have been held, to constitute or to display a mental and theological pattern distinctive from the thinking of other peoples. This distinctiveness appears very strikingly when the comparison made is with the patterns of Greek thought. Thus an interest in the unity and the distinctiveness of the Old Testament has been widely linked with such lexical procedures.

The same emphasis succeeds also in bringing the New Testament into the same picture. It is held that the thought of the Greek Bible, and especially of the New Testament, is emphatically not Greek thought but Hebrew thought, and that the Greek words of the Bible have thus received 'new content'. The thought pattern behind them is the Hebraic, and for the understanding of them nothing is more important than to see how these Greek words filled with this new content stand in strong opposition to their uses within pagan Greek thought. Once again, then, the assembly by an appropriate procedure of the significant theological terms of the Greek New Testament brings out not only its own inner coherence but also its full coherence with the Old Testament, and not only this but its strong discontinuity with the life of the Hellenic mind and thinking. To demonstrate this was explicitly the main purpose of the large and still unfinished Kittel dictionary.[1]

We see then that an emphasis on a lexical approach has in fact been commonly linked with the prevailing assertions of the coherence and distinctiveness of the Bible. The general difficulties and faults in the lexical approach thus made have been dealt with by the author elsewhere,[2] and will not be repeated here, for here only one particular procedure within the lexical field is to be investigated. Nor will the attempts[3] to do the same with non-lexical aspects of biblical language, such as the organization of the Hebrew verb system, be mentioned here except incidentally. I must reiterate that my criticism of this lexical approach does not intend or imply any attack on the idea of the coherence or distinctiveness of the Bible, except where such coherence or

---

[1] On this see *Semantics*, p. 241 f. and generally.
[2] *Semantics*, pp. 107-60 and 206-62.
[3] In particular by T. Boman in his *Hebrew Thought compared with Greek*.

distinctiveness is understood in a sense inextricable from involvement in the lexical procedures which have sometimes been allied with it.

The ultimate linguistic fact, to which the lexical arguments for the coherence of biblical thought appeal (perhaps unconsciously), is the existence of one great language system for the entire Old Testament, namely Biblical Hebrew (we may leave aside the Aramaic sections for the present), and one for the New, namely New Testament Greek. While there will be differences in style and in choice of words between a writer of David's time and Isaiah, it remains intelligible and reasonable to assign them alike to 'Biblical Hebrew'. Paul and Mark wrote at about the same time. Their words and styles differed, and one can speak if one wishes of 'Paul's language' as something other than 'Mark's language'. But in another sense, and the one I am using here, Paul and Mark 'spoke the same language'. It was one system, though they used its resources differently; its grammar can be described more or less as one piece, and its lexical stock can be inventoried. This fact makes possible the common theological approach through lexical method. Basically this approach overcomes the variety of date, emphasis and interest among the biblical documents by emphasizing the community of the lexical stock. To put it conversely, this is something that could not have been done had each of the prophets, or each of the apostles, spoken a different language, so that the original text of our Bible would be in twenty or forty different tongues.

I have already stated the nature of the procedure which is to be investigated in this study. It is the building of a structure from the lexical stock of the biblical languages, and with it the assumption that the shape of this structure reflects or sets forth the outlines of biblical thinking about the subject. This description is so far a loose one, and several procedures which differ in detail may be considered as the building of a structure on a lexical basis of some kind.

One form of such structure would be where a word is used with two senses, and this fact is taken to display or reflect a relation in which the objects or actions designated stand to one another. A good example is the use made in *TWNT* of the double sense 'comfort' and 'exhort' of the verbs παρακαλέω and παραμυθέομαι.[1] The fact that the same word spans both meanings is supposed to express or to reflect an actual relation of two elements in Christian life or thinking. Similarly, I have suggested that one might equally suppose the senses of Hebr. *niham* in two different themes, namely 'repent' (*niph'al*) and 'comfort' (*pi'el*), to express or reflect a real structure or relation between these actions in the mentality of Old Testament theological thinking.[2]

Again, attempts have been made to demonstrate, from a number of different words used to designate a single reality, the relation existing in fact between different aspects of that reality. Thus E. Jacob, for example, in an interpretation which I have already criticized elsewhere,[3] assembles a series of Hebrew words for 'man', and holds that each of these words typifies or emphasizes or brings out a particular aspect of the Hebrew understanding of man. Thus one word means 'man as earthly', another 'man as strong', another 'man as possessed of the power of choice', and so on. In this case my previous criticism rather concentrated on the way in which this procedure depended upon etymological considerations and neglected the semantics of actual usage. But even apart from this criticism, there is reason to doubt whether the procedure is a valid one. Where synonyms exist in a language, it is precarious to assume that the existence of more than one word for a reality must mean, or must lead to the recognition of, the existence of various aspects or emphases of that reality, each of which is indicated by one word.

Thirdly, we have attempts to construct a lexical structure from two or three words which are fairly close in meaning or which commonly relate to the same subject or theme. Such is in fact the procedure in the investigations of the biblical idea of time which this work will criticize. In the two chief studies of the New Testament conception which will be criticized here, the lexical material is in each case arranged so as to present a structure, in which two Greek words correspond to two chief components in the conception of time, and the conception of time is dominated by the pattern of this two-dimensional structure. In Marsh's work the two words upon which the structure is formed are

---

[1] *TWNT* v. 777 f., 819; *Semantics*, p. 232 f.
[2] *Semantics*, pp. 233, 265.
[3] *Theology of the Old Testament*, p. 156 f.; criticisms by the writer in *JSS* v (1960) 166 ff., and *Semantics*, pp. 144-7.

καιρός and χρόνος; in Cullmann's the words are καιρός and αἰών.

This procedure has been tried as a means to state the nature of biblical thinking about other subjects than time. Thus the attempt sometimes made to correlate three Hebrew words in the area of 'redeem, reconcile' with three aspects or movements in the actual divine work of redemption is the same procedure; it may try to correlate *ga'al* with a divine movement towards man, *pada* with a substitutionary movement within mankind, and *kipper* with the reciprocity of reconciliation, or something of the kind. Again, a *TWNT* article notices the number of Hebrew words for anger, and maintains that each of them will surely stand for a particular essential feature in the Hebrew understanding of anger, a subject important for the anger of God.[1] Such other attempts will not be further discussed here; my purpose in mentioning them is only to show that the procedure I criticize for time could be used also for other biblical conceptions.

In approaching the examination of the lexical arguments concerning time, certain points in method must be made clear.

Firstly, my criticism concentrates on the lexical procedures within the works studied, and makes no attempt to deal with other aspects of their argument, or of the theology from which they begin or with which they end, except in so far as these other aspects are necessarily linked with the lexical method. Large areas in the books of Marsh and Cullmann will thus not be commented on or criticized. On the other hand it is in the writer's opinion clear that the lexical structure in these works is entirely fundamental to the way in which the general case is presented. Moreover, it is a very important element in the conviction which the general argument carries, for the lexical structure makes the general conclusions appear to emerge inevitably from the very fibre of New Testament language; and this impression may inhibit, especially in the minds of students, criticisms and objections which might otherwise be raised. But I repeat that the general theological position of Marsh or of Cullmann is not under criticism here. The writer in fact belongs to much the same general theological tradition as they do. The criticism is of the method by which a lexical structure is used to demonstrate

[1] *TWNT* v. 392. See *Semantics*, p. 147 f.

certain conclusions, of the view of time reached by these methods, and also, by implication, of other theological results reached by these methods or by others involving similar presuppositions.

Secondly, a good deal of the argument consists in establishing points at which the linguistic evidence has been either mis-stated or seriously neglected. I must point out that this has been done only where such mis-statement or neglect arises, in my judgement, from the nature of the lexical method employed, and therefore stands as evidence for the faultiness of that method. I have not collected mistakes at random, or taken notice of any which are not connected with the lexical method criticized. There was, in the writer's submission, no other way to handle the problem. Readers would not wish, surely, to have an ideal lexical method expounded to them in theory, and then to be told that well-known positions such as those of Marsh and Cullmann were defective since they did not follow the theory. If it is demonstrated, however, that these well-known positions depend on a neglect of evidence which was already widely available, or on methods which are recognizably faulty even to those without technical or theoretical linguistic training, then readers may find themselves considering other methods in practice, whatever theory may have to say.

Thirdly, it is only to a limited degree that my argument produces, or depends upon, entirely new or revolutionary methods in the handling of linguistic evidence. It is in fact one of the advantages of the more negative approach, namely, that through criticism of earlier solutions, that faults in these solutions can be demonstrated on the basis of known and acknowledged evidence and methods, without appealing to methods which may indeed be better but which are unfamiliar in theological study and untried in the biblical area. By far the largest part of the evidence which I use in criticizing the lexical structures investigated is, and long has been, available in the standard works of reference. I do on the other hand believe that some new methods are much needed, and in chapter five I assemble lexical material in a way which is intended on the one hand to display the nature of the distribution of lexical stocks in general, and on the other hand to adumbrate (in the Greek and Hebrew sections) some lines of handling lexical material which might be developed farther subsequently.

## II

## THE *ΚΑΙΡΟΣ-ΧΡΟΝΟΣ* DISTINCTION
## —MARSH AND ROBINSON

In a full-length study of biblical ideas about time[1] Marsh lays
down a distinction between two chief possible ways of thinking
about time, the 'chronological' and the 'realistic'. The same
difference, put in another way, is that 'between time as chronological and time as opportunity'. The chronological mode is a
way of thinking of time merely as something measured, and with
no reference to what happens in the time so measured. The
realistic mode means thinking of the actions in time and of their
opportune occurrence; it has little or no interest in time in itself
as something apart from these events. Thus in the latter case
time is defined and known 'by its content'; 'times are known and
distinguished not so much by their place in some temporal
sequence as by their content; i.e. they are known realistically,
rather than chronologically'.[2]

Before we discuss some further points about this distinction,
we may pass on to what is for us by far the most important next
step, namely the exact correlation by Marsh of this distinction
with the difference between two words in the vocabulary of
New Testament Greek:

'The New Testament can reproduce our modern distinction.
Χρόνος is its word for chronological time, and the word
καιρός stands for realistic time'.[3] Or again: 'The contrast
between "chronological" and "realistic" is exhibited in the
NT distinction between χρόνος (measured time, duration) and
καιρός (time of opportunity and fulfilment)'.[4]

A rather similar procedure is followed by Robinson.[5] He
discerns 'a fundamental difference between two ways of regarding . . . the whole of the time process' and says that 'it is a
difference which the biblical writers indicate by their use of the

---

[1] J. Marsh, *The Fulness of Time* (1952), p. 19 f.
[2] Ibid., p. 21.  [3] Ibid., p. 20.
[4] Article 'Time' in A. Richardson, *A Theological Word Book of the Bible* (1950),
p. 258.
[5] J. A. T. Robinson, *In the End, God . . .* (1950), p. 45 ff.

two Greek words for time, καιρός and χρόνος'. He thus states the difference as follows:

'*Καιρός* is time considered in relation to personal action, in reference to ends to be achieved in it. *Χρόνος* is time abstracted from such a relation, time, as it were, that ticks on objectively and impersonally, whether anything is happening or not; it is time measured by the chronometer, not by purpose, momentary rather than momentous'. Or again: 'It is possible to abstract from the personal purpose and regard history simply as chronology or chronicle. *Χρόνος* is time removed from reference to God and regarded as self-determining. It is time given an autonomy which it does not by right possess.'

Now there are considerable further ramifications in the discussion of the matter by Marsh and Robinson, but before examining them we may set out the central criticism, in which I do no more than follow a valuable note by Professor Caird.[1] Even if we admit for the present that the contrast between their two modes of thinking about time is a valid and relevant one, the correlation of this contrast with the difference between καιρός and χρόνος in New Testament usage is wholly untenable. I set out below the three sets of passages cited by Caird:

(a)  i. Mark 1.15 πεπλήρωται ὁ καιρός
      'The time is fulfilled'.

    ii. Gal. 4.4 ὅτε δὲ ἦλθεν τὸ πλήρωμα τοῦ χρόνου
      'But when the fullness of time came . . .'

(b)  i. Acts 3.20 ὅπως ἂν ἔλθωσιν καιροὶ ἀναψύξεως
      'that times of relief may come . . .'

    ii. Acts 3.21 (part of the same sentence as the above) ὃν δεῖ
      οὐρανὸν μὲν δέξασθαι ἄχρι χρόνων ἀποκαταστάσεως πάντων
      'whom heaven must receive until the time of restoration of all'.

---

[1] G. B. Caird, *The Apostolic Age* (1955), p. 184 n. 2. Cullmann also seems implicitly but rightly to deny the validity of the καιρός-χρόνος distinction when he says that in the NT chronos appears 'more or less in the sense of kairos'; although, as we shall see, it is more important that, to speak in Cullmann's terms, 'kairos appears in the sense of chronos'. See Cullmann, *Christus und die Zeit*, p. 42; E.T., p. 49.

(c) i. I Peter 1.5 εἰς σωτηρίαν ἑτοίμην ἀποκαλυφθῆναι ἐν καιρῷ ἐσχάτῳ 'for salvation ready to be revealed in the last time'.

ii. I Peter 1.20 Χριστοῦ ... φανερωθέντος δὲ ἐπ' ἐσχάτου τῶν χρόνων 'Christ ... manifested in the last of times ...'

iii. Jude 18 ἐπ' ἐσχάτου τοῦ χρόνου ἔσονται ἐμπαῖκται

'In the last time there shall be scoffers ...'

It is surely wholly impossible that Mark 1.15 was talking about 'realistic time' and Paul in Gal. 4.4 talking about 'chronological time', or that χρόνος in any of these cases is 'time abstracted from relation to personal action' or 'time removed from reference to God and regarded as self-determining'. Caird's note needs no reinforcement from further examples; some other points raised in it, such as LXX usage, will be taken up later; for the present we register simply that it entirely explodes the correlation of the difference between χρόνος and καιρός with the distinction of 'chronological' and 'realistic' time.

We may stop here and summarize. (a) If there is a difference between χρόνος and καιρός in New Testament usage, it is clear that it cannot correspond to the distinction between 'chronological' and 'realistic' time as expounded by Marsh and Robinson. We shall return later in this chapter to discuss more fully the meaning of these words in biblical usage and the difference between them. (b) The correlation of the two words roughly meaning 'time' with two fundamental conceptions of time is no isolated example, but is a procedure quite common in some kinds of recent theological interpretation, and one which to those who use it clearly seems entirely natural and well-founded; it has been stimulated in particular by the methods of Kittel's *TWNT*.[1] Our further discussion will be directed therefore not only to showing that the differentiation of χρόνος and καιρός by Marsh and Robinson is an erroneous one in fact, but to considering whether it is also wrong in principle, that is, whether it may not be a basically unsound and misleading way of handling language to attempt this correlation in the first place. Meanwhile we return to take up some general points in the case of Marsh and Robinson.

(a) Marsh states explicitly[2] that he draws his distinction of the

---

[1] See the discussion of *TWNT* in my *Semantics*, pp. 206-62, and particular examples such as the handling of λόγος and μῦθος, p. 220 ff.

[2] *Fulness*, p. 19 ff.; *Word Book* 'Time', p. 258.

since there is nothing of 'time of opportunity and fulfilment' in the sentence except through very strained interpretation of it. Again, both of the contrasted views in iv. A and iv. B involve some kind of 'time known by its content', for a quite simple list of occurrences in time constitutes a statement of the 'content' of time, even if nothing more than accident is held to connect the elements of the series. The contrast of iv. A and iv. B does not coincide with that of i. A and i. B.

Again, in iii. A it may be possible in some sense to regard history 'merely as chronology' without interest in the 'content', i.e. the events (although one finds it hard to imagine how this can be done); but it is surely impossible to regard history 'merely as chronicle' without such an interest; so that iii. A in part if not entirely would seem to fit with i. B rather than with i. A. On the other hand, it would not seem impossible that some one might think of 'time as known by its content' (i. B), but at the same time regard time as something possessing some existence and power of its own (v. A). Again, the two elements mentioned in ii. B are hardly homogeneous, for opportunity and fulfilment are by no means the same in the respects here under discussion; though one may, it is possible, not think of 'opportunity' with reference to a duration, one may certainly so think of 'fulfilment', and there is good reason to suppose that the biblical writers did so.

One must say, then, that the distinction between 'chronological time' and 'realistic time' is by no means used or worked out clearly or consistently by either Marsh or Robinson. Moreover, certain important cases are not considered at all. For example, when we measure and state the time elapsing *between* two events, is this mere duration ('chronological time'), or does the reference to the two terminal events qualify it as 'realistic time'? This is of course important for many chronological statements in the Bible, to which we shall return shortly.

It also must be stated here in a preliminary way that the translations given of καιρός[1] contradict partly one another and partly also the contrasts in modes of thought about time. Thus Marsh and Robinson, following usual English practice, frequently translate καιρός by 'time'; but 'opportunity' also appears

---

[1] A fuller discussion of the importance and the nature of translation is given later; see pp. 110-16.

frequently. Discussing the phrase οἱ χρόνοι καὶ οἱ καιροί (I Thess. 5.1; Acts 1.7) Marsh[1] states that 'καιρός, as we have seen, is an opportunity', although this is surely an impossible rendering if used in the translation of the phrase under discussion; and when he comes to discuss and translate Acts 1.7 he expressly renders καιρούς as 'the various epochs of history', while at I Thess. 5.1 he goes back to the traditional English rendering of 'seasons'. Not only does 'epochs of history' fail to fit with 'opportunity', but it certainly involves 'duration' of some kind, which has already been classified as belonging to 'chronological time' (ii. A).

(c) One of the weakest points in the whole discussion by Marsh and Robinson is their depreciation of the importance of chronology in the Bible. Both of them admit that chronology, like 'chronological time', has a place in the biblical literature, but they claim that it is less important than 'realistic time'. Marsh says, as we have seen, that both modes of thought about time are found in the Bible as in our modern speech; but he thinks that their relative importance is reversed, so that, while we pay more attention to the chronological, the Bible says very little about it and concentrates its interest and its contribution within the areas of 'realistic time'.[2] Robinson says[3]

> 'The usual word for "time" in secular Greek is χρόνος. In the biblical writings the normal term is καιρός. It is normal, not merely in the sense that it is more common, but in that it represents the norm or proper standpoint from which time is to be understood. Καιρός is time considered in relation to personal action . . .'

Later he states that, where an excessive interest is shown in the calendar as a means of predicting the end of the world, this is an effect of 'the understanding of time as χρόνος', and he holds that 'according to the view of time as καιρός the calendar date is both irrelevant and unpredictable'.[4] This is not far from Marsh's position. Both writers assert that chronology has a comparatively small and unimportant place in the Bible.

To this it must be said that the Old Testament evidence

---

[1] *Fulness*, p. 117 ff.       [2] Ibid., p. 19 f.
[3] Robinson, p. 45 f.       [4] Ibid., p. 47.

points strongly in the opposite direction. The depreciation of the importance of chronology has already been noticed by Eichrodt as one of the chief weaknesses in the approach of Marsh.[1] It is perhaps true that many assessments of Old Testament theology in modern times have set little emphasis upon this evidence, but the evidence itself was always very plainly there, and interest in it is growing again.[2]

The Old Testament contains a complete and carefully worked-out chronological system, by which a large number of the important events (and a good many details of less apparent significance) can be dated in relation to one another, and in particular dated from the absolute datum point of the creation of the world. That there are certain discrepancies in the details, that different schemes appear in different textual traditions (as in the LXX or the Samaritan text), and that some of the data are legendary, do not alter the fact that it is a chronological system, and the centre of the greatest interest to the biblical writers. Such figures as the 430 years of the sojourn in Egypt (Ex. 12.40), or the 480 years from the Exodus to the beginning of the construction of Solomon's temple, are cardinal points in the scheme.

The details and the purposes of such schemes do not concern us here, but in general it may be suggested that a major interest was the provision of significant figures (in relation to the creation of the world, or to the total time for which the world would last) for great events in religious tradition such as the Deluge, the Exodus, and the construction of major sanctuaries. Moreover, at least in the later stages of the development of the scheme, and possibly as late as the second century B.C., the possibility that such schemes might assist in the calculation of the end of the world was certainly an influence. The widespread interest in such schemes is shown also by the Book of Jubilees, while the

---

[1] *ThZ* xii (1956) 113 ff. Eichrodt's criticisms apply to Boman and Ratschow as well as to Marsh. The present writer had reached, independently of Eichrodt, the view that this is a basic weakness of Marsh's position; and would wish to express it rather more strongly than Eichrodt.

[2] On the chronological system see A. Murtonen, 'The Chronology of the Old Testament', in *StTheol* viii (1954) 133-7; R. Pfeiffer, *Introduction to the Old Testament* (New York, 1941), p. 200 ff., with citation of earlier literature; Oppert, 'Chronology (I)', *Jewish Encyclopaedia* (London and New York, 1901-6), iv 64-68; L. Koehler, *Old Testament Theology* (London, 1957), p. 87 f., who says that the system is so designed 'that at any point in the process one can ask when the end and the fulfilment will come'; E. Nielsen, *Oral Tradition* (London, 1954), p. 103.

Jewish historian Demetrius, writing in Greek, had at his disposal, or produced, even more detailed chronological information.[1]

In the New Testament a somewhat similar scheme appears in Matthew ch. 1, although the numbers of years are not given, but only the generations. St Paul also very probably had an exact knowledge of these schemes, for in Gal. 3.17 he quotes the figures precisely as they were transmitted in the Greek and Samaritan tradition. Again, the explicit statement in the speech of Stephen that Abraham migrated into Canaan 'after his father had died' may be no mere slip. While it is possible that this might arise from a reading of Genesis with no chronological or arithmetical interest,[2] it is considerably more probable that it arose from the conscious following of a chronology similar in this respect to that found in the Samaritan text.[3] Philo not only gives the same interpretation, but appears to regard it as a matter of common knowledge.[4]

The existence of this fairly comprehensive chronological scheme, and of widespread and meticulous interest in it, in late Judaism and early Christianity is almost entirely neglected by Marsh and Robinson. While it is true that not all events are dated by the system, and one can quote chronologically vague datings like 'two years after the earthquake',[5] the existence of events not dated by the system cannot entitle us to depreciate its great importance, or to maintain that 'it is typical of Scripture not to locate an event by defining its place on a chronological

---

[1] See the fragments in Jacoby, *Die Fragmente der griechischen Historiker* iii. C. 666-71. Eupolemus had the same interest, and had reckoned the number of years from his own time back to Adam and back to the Exodus; see Jacoby, op. cit., p. 677.

[2] This is the suggestion of Jackson and Lake, iv. 70. But in fact the Stephen speech shows considerable interest in figures and chronology, e.g. Acts 7.6, 8, 14, 20, 23. Moreover, the suggestion of Jackson and Lake assumes that Stephen (or St Luke) read Genesis in a 'natural' way, in which the events of the passage Gen. 11.27-12.5 would be taken to have occurred in the order in which they are narrated. But we can prove beyond question that he did not, because he explicitly places the speech of Gen. 12.1 ff. in the period before the Terachids left Ur for Haran. It is more probable that Stephen had a chronological fixation for the call of Abraham, or for the departure from Ur. The best guess might be that he fixed it at the 60th year of Abraham's life, more or less as in Jubilees 12.12.

[3] By the figures in the MT Abraham migrated into Canaan in A.M. 2021, and his father Terah died 60 years later in 2081. In the Samaritan text the migration took place in A.M. 2322, and Terah's death took place in the same year. Cf. P. Kahle, *The Cairo Geniza* (1st ed., London, 1947), p. 143 f.; the 2nd ed. seems to omit the passage.

[4] *De migr. Abr.* 177. Cf. Jos. *Ant.* i, 152.        [5] Marsh, *Fulness*, p. 21.

scale.'[1] The chronological scheme is of fundamental importance for the understanding of the biblical text.[2]

Marsh, while he grants the existence of some kind of chronology within the Old Testament, also minimizes it with several further arguments. Fist of all, he argues that Hebrew has no word which can translate χρόνος or 'chronological time', while it has a wealth of words for 'realistic time'; this argument will be discussed later.

Secondly, he maintains 'that chronology (as we know it) was a late achievement of the Hebrew mind, and that, when it was developed, it was largely borrowed'.[3] Against this one must say: (a) While it is likely to be true that the full chronological system of the Bible is a late development, some at least of the elements in it, such as the date of 480 years at I Kings 6.1, are most probably quite ancient. (b) A large part of the chronological system, or of some earlier version of it, is in all probability integral to the final redaction of the Pentateuch and the historical books, forming as it does a chief *raison d'être* for not a few passages in the framework of this literature; so that to cast doubts on the importance of the chronological scheme because of its late date is also to question the whole process by which the Old Testament reached its final shape. (c) Even if it is true that the chronological interest is 'borrowed', this can hardly mean it is invalid as evidence for the Hebrew understanding of time, since it was the Hebrews themselves presumably who voluntarily 'borrowed' and developed it; and Marsh himself quotes as 'characteristic of the whole biblical tradition' a calendar system which he himself explicitly declares to be 'probably Canaanitish'.

Thirdly, we have Marsh's argument from the names of the months in the various Hebrew calendars. The latest calendar uses the Babylonian names like Nisan or Tishri. The point intended is that these are more or less just labels, which give no idea of the activity-content of the month named. Marsh says that some of the series are not used by the biblical writers; but if this is meant to suggest that these writers had a distaste for names

---

[1] Marsh, *Fulness*, p. 20 f.

[2] It may be that apologetic reasons have been influential in causing the neglect of the scheme; for naturally it is not easy to persuade modern people to take seriously some of the biblical figures in the literal sense in which they are clearly intended.

[3] *Word Book*, 'Time', p. 258.

so lacking in 'time defined by its content', this must be pronounced a most unlikely deduction. An earlier series designated
the months by their order, e.g. 'the first month', etc. Still earlier
we have the system of Canaanite origin 'in which the months
were named by what happened in them', giving us Abib 'month
of ripening ears', Ziv 'month of flowers', etc. Of this system
Marsh says 'this designation of a "time" by its content is characteristic of the whole biblical tradition and is connected with the
understanding of time "realistically" in terms of opportunity
and fulfilment'. But the serious weaknesses in this argument are
very obvious. There are only a handful of passages where these
names are used in the Bible, and only four names of the series
have survived, three of them used only once or twice. How can
we say that a system which is hardly used at all in the Bible is
'characteristic of the whole biblical tradition'? The only significant
contribution the Israelites made to the designation of time (as far
as concerns the names of months) seems to be that, having
borrowed this system from the Canaanites, they soon forgot it in
order to adopt another which was less characteristic of biblical
thought about time!

Before leaving the subject of chronology we may add something about an argument of von Rad.[1] He believes that the
Israelite way of thinking about time is wholly different from our
own, and that earlier exegetes naïvely assumed their own modern
idea of time to be valid for Israel. Israel knew only 'filled time';
time is known only from its own content. Now the books of
Kings offer chronological schemes of the kings of Judah and
Israel, and these are, in von Rad's opinion, clearly independent
streams of time, each filled with its own content; one time is
filled with the being of the Judaean, the other with that of the
Israelite kings. To us it would seem natural to bring these two
streams together and incorporate them in one line; but this step
was never taken; 'each of the two series of kings contains its own
time'.

But surely this is unconvincing, and shows how scholars,
having decided that a different idea of time existed in Israel, are
forced to find most unlikely supports for it. Certainly each of
the two kingdoms reckoned its royal chronology for itself, but

[1] *Theologie des alten Testaments*, ii. 112 ff.

there is not the slightest atom of evidence that this should be interpreted as a conception that two wholly independent streams of time were involved. Von Rad says that no attempt was made to bring the two streams together, but what else is effected by the synchronisms, which are regularly provided, between the regnal years of the kings of the two countries? There is not the slightest reason to deduce that two separate streams of time were involved. If each chronology had been reckoned from a datum point of its own, for instance the accession of David for the Judaean and that of Jeroboam for the Israelite, and no reference had been made to the reigns of the other kingdom, one might have had some ground for thinking of separate streams of time. But this is just what was not done, and in fact synchronisms are regularly given. Naturally, as von Rad says, we must be open to the possibility that another idea of time than our own existed in Israel, but the production of interpretations like this one does not encourage us to hope that such openness will necessarily lead to sensible and natural explanations of the facts.

We sum up this section then by saying that Marsh's statement that 'it is typical of Scripture not to locate an event by defining its place on a chronological scale, but to identify it by its content' is far from doing justice to the facts. The chronological system is of fundamental importance for the Old Testament.

Our results up to date then are: (a) following Caird, we see that, if there is any difference between χρόνος and καιρός in New Testament usage, it cannot be correlated with the difference between 'chronological time' and 'realistic time', and that in some passages of central theological significance there may be good reason to suppose that there is no real difference between the words; (b) the difference between 'chronological time' and 'realistic time' appears to contain inner contradictions which make it unsuitable as an instrument to guide us in the study of the meaning of words; (c) there is no good ground for considering chronological reckoning to be an unessential or secondary element in the biblical understanding of time.

With these results in mind we can now make a fresh approach to the meaning of καιρός and the question of its opposition to χρόνος. The following study will not try to give a complete account of the history of the words, but to bring out certain aspects of usage which are relevant to the understanding of time in the New Testament and which have been neglected in presentations such as those of Marsh and Robinson.

We begin by simply registering the well-known classical usage of καιρός in the sense of 'exact, right, critical time' or 'opportunity'. This is accompanied by senses of 'due measure' or 'right proportion' in no reference to time, a case therefore which is of no direct relevance to our subject. Where καιρός has a reference to time in a classical author like Aeschylus[1] the sense is roughly that of 'opportune time'.

It is however clearly stated by LS and other authorities that the sense of 'generally, *time, period*' (so LS) begins to appear for καιρός after classical times. In the plural καιροί this sense appears as early as Aristotle's *Athenian Constitution*. In this work at xxiii. 2-3 we find: καὶ ἐπολιτεύθησαν Ἀθηναῖοι καλῶς κατὰ τούτους τοὺς καιρούς· συνέβη γὰρ αὐτοῖς κατὰ τὸν χρόνον τοῦτον τά τε εἰς τὸν πόλεμον ἀσκῆσαι . . . ἦσαν δὲ προστάται τοῦ δήμου κατὰ τούτους τοὺς καιροὺς Ἀριστείδης . . . καὶ Θεμιστοκλῆς . . . While it might be possible to understand this as 'at those critical times (of the Persian wars)', and likewise at xxxiii. 2,[2] some other passages seem strongly to support the general sense of 'time' and to suggest that LS are right in giving it as the sense for xxiii. 2. Thus at xli. 1 we read ταῦτα μὲν οὖν ἐν τοῖς ὕστερον συνέβη γενέσθαι καιροῖς, τότε δέ 'these things came about in later times, but then . . .', and at xvi. 10 ἦσαν δὲ καὶ τοῖς Ἀθηναίοις οἱ περὶ τῶν τυράννων νόμοι πρᾶοι κατ' ἐκείνους τοὺς καιρούς. Cf. also xxii. 7. This usage of the plural continues later, and is found for example in Polybius; at v. 33. 5 τὰ κατὰ καιρούς means roughly 'events from time to time'.

For the singular καιρός the general sense of 'time' or 'period' is particularly common in Polybius, and the following examples very probably show interchangeability of καιρός, καιροί and χρόνος: xxi. 40. 1: κατὰ τοὺς καιροὺς τούτους κατὰ τὴν Ἀσίαν Γναΐου . . . παραχειμάζοντος ἐν Ἐφέσῳ . . . παρεγένοντο πρεσβεῖαι . . . '. . . at this period embassies arrived . . .'

---

[1] Italie, *Index Aeschyleus* (1955), s.v.
[2] At xxxiii. 2 H. Rackham in the Loeb edition gives the translation 'during this critical period'.

xxi. 41. 6: κατὰ δὲ τὸν καιρὸν τοῦτον οἱ δέκα πρεσβευταὶ εἰς "Εφεσον κατέπλευσαν
'at this time the ten legates arrived by sea at Ephesus . . .'

xxvii. 1. 1: ἐν τῷ καιρῷ τούτῳ παρεγένοντο πρέσβεις . . .
'at this time there arrived as envoys . . .'

xxvii. 1. 7: κατὰ δὲ τὸν καιρὸν τοῦτον ἐν ταῖς Θήβαις συνέβαινε ταραχὰς εἶναι καὶ στάσεις
'at the same period there were quarrels and disturbances at Thebes . . .'

xxi. 6. 2: κατὰ τοὺς αὐτοὺς χρόνους . . . ἐξέπεμψαν πρεσβευτὰς πρὸς Σέλευκον . . .
'at the same date . . . they sent envoys to Seleucus . . .'[1]

It may well be, however, that in certain contexts only χρόνος is used, for example such contexts as 'after the lapse of time', e.g. xxi. 18. 8: χρόνου δ' ἐγγινομένου ὁ μὲν βασιλεὺς ἐξεχώρησεν . . .

One more case from Hellenistic literature, and still nearer to the New Testament period, may be given. Strabo in xvii. 46 (C 816) tells of how a maiden of the Egyptian Thebes was required to undergo a period of compulsory prostitution, and goes on: μετὰ δὲ τὴν κάθαρσιν δίδοται πρὸς ἄνδρα· πρὶν δὲ δοθῆναι, πένθος αὐτῆς ἄγεται μετὰ τὸν τῆς παλλακείας καιρόν. Surely καιρός can only mean a period of time.

An example containing both καιρός and χρόνος will be found in the testament of Strato of Lampsacus, who writes: τὸ δὲ περιὸν ἀργύριον κομισάσθω 'Αρκεσίλαος παρ' 'Ολυμπίχου, μηθὲν ἐνοχλῶν αὐτὸν κατὰ τοὺς καιροὺς καὶ τοὺς χρόνους, perhaps 'Let Arkesilaos obtain the rest of the money from Olympichus, but without bothering him in respect of the time and date'.[2]

For inscriptional evidence we may cite the Athenian inscription of 302/1 B.C., in which it is resolved to honour certain persons who διατετελέκασι ἐν παντὶ τῶι καιρῶι εὔνους ὄντες τῶι δήμωι τῶν 'Αθηναίων. Other similar cases can be quoted.[3]

Similarly for the papyri Preisigke gives the sense 'course of

[1] According to Dieterich in *RhM* lix (1904) 233 ff., Polybius represents the first entry of this sense into literary sources (presumably making a separate case of the plurals in Aristotle); he also quotes from Herodian a case of εὔκαιρος καιρός, which he sees as a clear indication of how much of the older characteristic sense of καιρός has been lost.

[2] F. Wehrli, *Die Schule des Aristoteles*, v. (Basel, 1950), frag. 10:32 f.

[3] Ditt. *Syll.* 346, 10; cf. 374, 9 and 443, 20.

time' as well as those of 'point of time' and 'opportune time'. As early as the third century B.C. κατὰ τοὺς αὐτοὺς καιρούς can be quoted in the sense of 'at the same time', and καθ' ὃν καιρόν can mean simply 'when'.[1] Again, in a papyrus of the third century B.C., of a time therefore of prime importance for the LXX, we find the phrase τὸν δὲ καιρὸν ἐκτρέχειν, where in the context the use of the present infinitive and other factors suggest that the sense is 'that the time (available) is running out' rather than 'that the critical moment is passing'.[2]

On the other hand χρόνος is also common, and indeed much commoner, being used in the epistolary style with the sense of 'date', as in ὁ χρόνος αὐτός 'of the same date', in addition to other usages. The study of καιρός by Moulton and Milligan begins by citing passages 'for the idea of "fitting season", "opportunity", which is specially associated with this word', and although they go on to note certain other senses, such as those near to 'crisis' or to 'prosperity', and the transition to the sense of 'weather' existing in Modern Greek, they give no indication at all that the more general sense of 'time' was current and familiar in Hellenistic usage, and would need to be considered possible, to say the least, for the New Testament passages.[3] The case they quote for the conjunction of χρόνος and καιρός, and therefore as a possible parallel to Acts 1.7 and I Thess. 5.1, with the idea that the two words 'bring out respectively the period and the occurrences by which it was marked', is hardly a good parallel to these NT passages; for in μὴ ὅτι γε τοσούτου χρόνου ἐπιγεγονότος καὶ τοιούτων καιρῶν 'to say nothing of so long time having passed, and such times!', the understanding is considerably affected by the τοιούτων, to which there is nothing corresponding in the NT passages.[4]

Recognizing therefore that there is a large area of post-classical Greek usage in which καιρός has a sense not of 'right time' but rather of 'time', we may approach the LXX usage. Here Caird in the note quoted states that there is no hard and fast distinction between καιρός and χρόνος in the sense suggested by Marsh and Robinson, and that generally καιρός means a point of time at

[1] *UPZ* 21, 8; cf. Ditt. *Syll.* 535, 5.
[2] The translation is given as 'que le temps passe' by M. Hombert in his publication of the text in *Revue belge de philologie et d'histoire* iv (1925) 654 f.
[3] MM, p. 315.　　　　[4] Ibid., p. 694.

which something happens, and χρόνος a period of time; but that even this distinction is not always maintained. In fact this point can be made even more strongly, as the evidence adduced below will indicate.

In the LXX the total cases of καιρός are more than three times more numerous than those of χρόνος. Variations occur between one book and another. Isaiah, for example, actually has more cases of χρόνος than of καιρός, but a good proportion of them (7 out of 18 or so) occur in special phrases such as *lᵉ-'olam* 'for ever' translated as εἰς τὸν αἰῶνα χρόνον. In Jeremiah it is well known that all cases of καιρός occur before the end of ch. 28 and all cases of χρόνος (4 in all) occur after that point, a circumstance which was duly noted by Thackeray in connection with the change of translation techniques in this book.[1] There are some books, such as the Dodecapropheton, which do not use χρόνος at all. In I Esdras on the other hand it is more frequent than καιρός. The greatest number of cases of καιρός in a single book is in Sirach (about 60, against 3 of χρόνος); but in Wisdom and II Maccabees we have roughly equal numbers of each word.

These general figures however tell us little, since the frequency with which the two Greek words are used has to be correlated with the frequency of various contexts within various books. We can begin by showing examples where χρόνος is used in contexts where either a moment of time or a decisive or critical time is meant, and where therefore according to the scheme of Marsh and Robinson we ought to expect καιρός. Pretty clear cases include:

(a) Dan. 4.34 (LXX version): Nebuchadnezzar relates that καὶ ἐπὶ συντελείᾳ τῶν ἑπτὰ ἐτῶν ὁ χρόνος μου τῆς ἀπολυτρώσεως ἦλθεν. 'at the end of the seven years the time of my release came'.

(b) Jer. 45 (38).28 καὶ ἐκάθισεν Ιερεμιας ἐν τῇ αὐλῇ τῆς φυλακῆς ἕως χρόνου οὗ συνελήμφθη Ιερουσαλημ. 'Jeremiah sat in the court of the prison until the time when Jerusalem was captured.'[2]

In addition we may cite one case where χρόνος appears to be used in a sense not only of 'time' with reference to a critical

[1] *The Septuagint and Jewish Worship* (London, 1921), p. 116.
[2] Cf. also such cases as I Esd. 6.3; Esth. 2.15.

moment but even of 'opportunity, occasion', that is to say, that sense in which καιρός is commonly much more distinct in our literature:

I Esd. 9.12 πάντες ... ὅσοι ἔχουσιν γυναῖκας ἀλλογενεῖς παραγενηθήτωσαν λαβόντες χρόνον.

'Let all those who have foreign wives appear when they have opportunity.'[1]

Similarly, considerable evidence is available in the LXX for καιρός meaning 'time' in cases where a period of time is intended. The most obvious is the famous Dan. 7.25 ἕως καιροῦ καὶ καιρῶν καὶ ἕως ἡμίσους καιροῦ 'for a time and times and half a time', with similar texts in both versions both here and at 12.7, where only periods of some kind can be meant. The same is true in the Theodotion version at 4.16 καὶ ἑπτὰ καιροὶ ἀλλαγήσονται ἐπ' αὐτόν 'seven periods shall pass over him'; cf. the LXX version's ἑπτὰ ἔτη here, and also Theodotion at 4.23, 25, 32.[2] Likewise in Ps. 70 (71).9 (MT *'al tašlikeni l°'et ziqna*) the sense 'moment' is surely impossible, and the sense 'time, period' required, by μὴ ἀπορρίψῃς με εἰς καιρὸν γήρους 'cast me not away in the time of old age'.

The same is most probable in a case like Amos 5.13 (cf. Micah 2.3) διὰ τοῦτο ὁ συνίων ἐν τῷ καιρῷ ἐκείνῳ σιωπήσεται, ὅτι καιρὸς πονηρός ἐστιν ... 'Therefore the discriminating shall be silent in that time, for it is a bad time'.

It is certainly true also of cases with the plural καιροί, such as Ezek. 12.27 ἡ ὅρασις ... εἰς ἡμέρας πολλάς, καὶ εἰς καιροὺς μακροὺς οὗτος προφητεύει, 'The vision is for many days, and for a long time (ahead) he prophesies', or I Macc. 12.10 πολλοὶ γὰρ καιροὶ διῆλθον ἀφ' οὗ ἀπεστείλατε πρὸς ἡμᾶς 'for a long time has passed since you wrote to us'.

The sense 'period, course of time' is also certain in II Macc. 4.17 ἀσεβεῖν γὰρ εἰς τοὺς θείους νόμους οὐ ῥᾴδιον, ἀλλὰ ταῦτα ὁ ἀκόλουθος καιρὸς δηλώσει 'for it is not easy to act impiously against the divine laws, but future time will disclose these things.'

Sufficient evidence has now been given to make clear that, as Caird has noted, the difference between χρόνος and καιρός in the

---

[1] This case appears to be without parallel in the LXX.
[2] In later Greek, however, it was not καιρός but χρόνος which took on the sense of 'year'; cf. Löfstedt, *Late Latin*, p. 117 f.; it may however be worth considering whether this passage from Daniel should not be cited as relevant to the discussion of that development.

LXX cannot be stated as that between 'chronological time' and 'realistic time', and that even the purely temporal distinction between a period of time and the point of time when something happens is not a correct statement of the difference between the words. Moreover, this is not all the evidence which can be furnished from the LXX, and some further points will be added later. Even in cases where it *might* be possible to argue that καιρός means 'decisive time', we must clearly understand that we are not entitled to feel that this sense has some kind of inherent probability, or some right of precedence over the other.

In general Marsh states about the LXX that[1]:

> 'It would be a gross mistake to suppose that when they (these words) appear in the New Testament they are just ordinary Greek words, for whose meaning an ordinary dictionary of Classical or even Hellenistic Greek must be exclusively authoritative. On the contrary, helpful and important as such dictionaries (especially an Hellenistic one) are, their definitions must be subordinated to those meanings discovered in the Septuagint translation . . . which embody the meanings with which the Hebrew words of the Old Testament are charged.'

But Marsh's own exposition depends throughout on emphasizing for καιρός a distinctive sense which is far more regulative in classical Greek than in the Septuagint, and which in fact appears only in fairly rare contexts in the latter.

At this point we may conveniently interpose some quotations from the Greek text of the Testaments of the Twelve Patriarchs, though without committing ourselves to an opinion of the date of this work in Greek. The following three passages deserve attention:

*i.* T. Sim. 2.6 ἐν γὰρ τῷ καιρῷ τῆς νεότητός μου πολλὰ ἐζήλωσα τὸν Ιωσηφ.

'In the time of my youth I was often jealous of Joseph.' The reference is undoubtedly to the whole period of his youth.

*ii.* T. Jud. 20.4 καὶ οὐκ ἐστὶ καιρὸς ἐν ᾧ δυνήσεται λαθεῖν ἀνθρώπων ἔργα.

'There is no time during which the deeds of men will be able

[1] *Fulness*, p. 76.

to pass unnoticed.' It is unlikely that the sense is 'critical time'.

*iii.* T. Dan 6.6 ἔσται δὲ ἐν καιρῷ τῆς ἀνομίας τοῦ Ἰσραηλ . . .

'But it shall be in the time of Israel's iniquity.' It is extremely probable that a period is intended, and not a 'critical time' or 'opportunity'. The phrase is almost certainly connected with the Qumran phrase *qeṣ ha-riš'a*.

It would be possible to add here also a study of the usage of still other relevant writers, such as Josephus; but this will be left aside here. The study of Josephus in this regard is complicated by the variety of his literary assistants, and will be much easier when the Josephus lexicon is finished. Let it suffice that specimen surveys suggest some degree of interchangeability of καιρός and χρόνος in certain contexts in Josephus as in biblical Greek. Thus in the same work and in closely neighbouring passages we have for 'at the time when' both καθ' ὃν καιρόν and καθ' ὃν χρόνον (*Vita* 3 and 13). In the *Antiquities*, for 'at that time' in historical narration we have κατ' αὐτὸν τὸν χρόνον at xix. 278 and κατὰ τοῦτον τὸν καιρόν at xx. 17. Many similar examples can be quoted.

We may now briefly establish the same point for the New Testament. Here most cases of χρόνος 'time' are used with reference to some kind of prolonged period, but in Acts 1.6 the question εἰ ἐν τῷ χρόνῳ τούτῳ ἀποκαθιστάνεις τὴν βασιλείαν τῷ Ἰσραηλ; 'are you restoring at this time the kingdom to Israel?' is not likely to mean 'in the course of this period' but rather simply 'at the present juncture', 'now', and could almost certainly be replaced by ἐν τῷ καιρῷ τούτῳ without change of sense; cf. the almost identical phrase at Esth. 4.14 ἐὰν παρακούσῃς ἐν τούτῳ τῷ καιρῷ 'if you pay no heed at this time . . .', and very numerous cases of ἐν τῷ καιρῷ ἐκείνῳ 'at that time'.

It is even easier to show examples where καιρός should be translated 'time' but has reference to what can only be an extended period. Thus for example:

Eph. 2.12 ἦτε τῷ καιρῷ ἐκείνῳ χωρὶς Χριστοῦ 'you were at that time without Christ'. So also Mark 10.30, parallel Luke 18.30 ἐὰν μὴ λάβῃ ἑκατονταπλασίονα νῦν ἐν τῷ καιρῷ τούτῳ 'except he receive a hundredfold now in this time', supplemented later by a statement of what will be received ἐν τῷ αἰῶνι τῷ ἐρχομένῳ 'in the coming age'. The man does not receive his rewards at this instant or moment, but in this time or age as contrasted with the coming period. In Heb. 9.9-10, though μεχρὶ καιροῦ διορθώσεως may perhaps refer to a moment of correction rather than a period, the description of something as a παραβολὴ εἰς τὸν καιρὸν τὸν ἐνεστηκότα can only mean 'a similitude referring to the present era'.

It is not necessary to multiply examples further. I wish the reader to remember however that I am not saying anything new here. The essential facts about the usage of καιρός have been plainly available for many years in a competent dictionary like Bauer. Even Delling's article in *TWNT*, which is in my opinion considerably biassed in favour of the sense of 'decisive time' for καιρός, registers ample evidence for the general senses of 'time' and 'period', although it rather obscures this evidence by putting it in small print. The method adopted by Marsh and Robinson, namely the setting up of two great conceptions of time and the correlating of these with the two familiar Greek words, simply ignores evidence amply collected and set out in the normal works of reference for the subject.

In particular, in the important passages (in the NT Acts 1.7 and I Thess. 5.1) where χρόνος and καιρός occur together in a collocation like περὶ δὲ τῶν χρόνων καὶ τῶν καιρῶν, we may be fairly sure that there is no significant difference between the two words, or at least none that can be expressed either by the distinction between 'chronological time' and 'realistic time' or by that between periods of time and moments of time. Such contrasts can scarcely be seen, as Caird points out, in those LXX passages where the two words occur and which can reasonably be supposed to have been in the background of the Acts passage. Thus compare:

Dan. 2. 21: *hu' mᵉhašne' 'iddanaya' wᵉzimnaya'*, where
Greek in both versions: αὐτὸς ἀλλοιοῖ καιροὺς καὶ χρόνους,
and Dan. 7.12: *'ad zᵉman wᵉ'iddan*, where
LXX: ἕως χρόνου καὶ καιροῦ
and Theodotion: ἕως καιροῦ καὶ καιροῦ

In view of the great frequency in Daniel, in the original Hebrew and Aramaic and in both Greek versions, of various words for 'time' in all sorts of combinations, such as ἕως συντελείας

καιροῦ (Theod., 9.27), or 8.17, where for Hebr. *lᵉ-'et qeṣ* we have
  LXX εἰς ὥραν καιροῦ
  Theod. εἰς καιροῦ πέρας,

it must be very doubtful whether all of these words are carefully
picked for various conceptions of time supposed to be clearly
indicated by each. It is more likely that the conjunction of
different words is not a representation of conceptions thus
distinct, but the familiar stylistic device of the juxtaposition of
several words where the matter referred to is one of notorious
vagueness or obscurity. In this case we can end with one other
example, where it is the prerogative of divine Wisdom to understand the outcome of times and periods:

εἰ δὲ καὶ πολυπειρίαν ποθεῖ τις
οἶδεν τὰ ἀρχαῖα καὶ τὰ μέλλοντα εἰκάζει . . .
σημεῖα καὶ τέρατα προγινώσκει
καὶ ἐκβάσεις καιρῶν καὶ χρόνων  (Wisd. 8.8).

For Acts 1.7 οὐχ ὑμῶν ἐστιν γνῶναι χρόνους ἢ καιρούς Blass
asserted that the two words are synonyms. Now Blass was one
of the greatest New Testament philologists who have ever lived;
yet Marsh bluntly denies his judgement here; and with it that
of the detailed work on Acts edited by Jackson and Lake, namely
that in such contexts the original distinction between the words
had been forgotten. The misuse of language in recent 'biblical
theology' is nowhere more evident than where the judgements
of careful scholarship are thus thrown aside because they do not
fit the lexical-theological web woven by recent fashions.[1]

We are now in a position to state tentatively the extent of the
lexical opposition between χρόνος and καιρός in biblical Greek.
The following statements must be taken as tentative and regarded with caution, for such reasons as the following. These
reasons are a consequence of our inability to examine the language
of the Greek-speaking writers of the OT and NT as a whole,
since we can work only from such evidence of their usage as has
in fact survived. (a) We may not only suspect, but know for
certain, that at some points the usage of one biblical book differs
from another; some illustrations from the translation of Hebrew

---

[1] Marsh, *Fulness*, p. 119, and citations there.

words in the LXX have already been given. (b) At points we may note that a certain combination never appears, and in this case we must beware of overloading an argument from silence so as to say that the usage is not a possible one; all we can say is that it does not appear, a statement the value of which is relative to the statistics of contexts where it might reasonably be expected to appear. (c) We have to be careful in judging at any point whether word A must mean something rather different from word B, or conversely that word A could be replaced by word B without any change of sense.

With these warnings in mind we may at least state some cases where χρόνος and καιρός seem to be opposed or to be interchangeable. Our main point is that in biblical usage these two words show oppositions[1] in certain syntactical contexts and none in others. It follows that the distribution of the words cannot be directly related, as is the principle of Marsh and Robinson's method, to distinct conceptions of time. The question whether they show opposition is syntactically determined.

Thus for example the opposition is obvious in:

ἐν καιρῷ 'at the right time, in season' (e.g. Matt. 24.45);
ἐν χρόνῳ 'in course of time' (e.g. Wisd. 14.16).

In this context the meaning of καιρός which was common in classical times is retained. The opposite is the case in:

ἐν τῷ καιρῷ ἐκείνῳ 'at that time';
ἐν τῷ χρόνῳ ἐκείνῳ 'at that time' (e.g. Tob. 2.11, S text);

this case has already been discussed above.

When defined by a genitive or possessive pronoun the commonest use is of καιρός, and the sense is 'the proper time' or 'the season' of something; e.g. Mark 11.13 ὁ γὰρ καιρὸς οὐκ ἦν σύκων 'for the time was not that of figs'. But in such contexts the duration of something is usually expressed by χρόνος, e.g. Dan. 4.27 LXX καὶ πλήρης ὁ χρόνος σου 'and your time is complete', or I Peter 1.17 τὸν τῆς παροικίας ὑμῶν χρόνον 'the time of your sojourning'. But there are some cases where χρόνος in such settings does not mean 'duration', as in Matt. 2.7 τὸν χρόνον τοῦ φαινομένου ἀστέρος 'the time of the star's appearing', or in Acts

---

[1] It should perhaps be explained that I use 'opposition' in the linguistic sense, of situations where a contrast or difference exists between two forms, and not in the more absolute sense, as in cases like 'black and white' or 'good and bad'.

7.17 ἤγγιζεν ὁ χρόνος τῆς ἐπαγγελίας 'the time of the promise drew near'. It is difficult in these cases to decide whether καιρός would not mean the same thing; it is most probable that it would.

With adjectives of quantity χρόνος is normal, e.g. μετὰ πολὺν χρόνον 'after much time' (Matt. 25.19), χρόνους ἱκανούς 'for a good time' (Luke 20.9), ἔτι χρόνον μικρόν (John 7.33) 'for a short time longer'; but occasionally καιρός is found, and the most notable example is at Rev. 12.12 ὀλίγον καιρὸν ἔχει 'he has little time'; this in a book which shows χρόνον μικρόν twice (6.11; 20.3). We have already seen πολλοὶ καιροί 'much time' at I Macc. 12.10. In this context, then, we must say that χρόνος is normal but that καιρός occurs occasionally.

This investigation is only a rudimentary one; much more could perhaps be done to classify the syntactical contexts in which καιρός means 'right time'. But taken along with the evidence already provided for the sense of 'moment', and for cases of theological importance where χρόνος and καιρός are almost certainly interchangeable, we may state with some assurance that in the LXX and NT καιρός keeps the special meaning, in which it shows opposition to χρόνος, of 'right time', only in certain contexts; and that over a large area of the usage, much larger than the number of the examples we have actually cited, the two words mean the same thing; but καιρός is in general a good deal more frequent. In particular, in those theologically important cases which speak of the 'time' or 'times' which God has appointed or promised, the two words are most probably of like meaning. Such cases of καιρός are, for example, Mark 1.15; Luke 21.24; Acts 1.7; 17.26; Eph. 1.10; I Thess. 5.1; I Tim. 2.6; 6.15; Titus 1.3; Heb. 9.10; while with χρόνος we have Acts 1.7; 3.21; 7.17; I Thess. 5.1.

One or two special cases may be mentioned here:

(a) The cases of καιροῖς ἰδίοις at I Tim. 2.6; 6.15; Titus 1.3, and of καιρῷ ἰδίῳ at Gal. 6.9 cannot be taken, as Cullmann[1] appears to intend, as if ἴδιος 'proper' were added in order to reinforce the sense of 'proper' already inherent in καιρός as 'right time'. Rather the reverse is the case. In the same writer καιρός does not mean 'right time' in ἐν ὑστέροις καιροῖς at I Tim. 4.1, and it is

---

[1] *Christus und die Zeit*, p. 34; E.T., p. 40.

rather the conjunction with ἴδιος, which has a similar effect to that with a genitive or possessive (as mentioned above), that fixes the sense of καιρός here as 'right or proper time'.[1]

(b) The case of ὁ καιρὸς συνεσταλμένος ἐστιν at I Cor. 7.29 is clear evidence for καιρός in the sense of 'time, period', although it has not been adduced up to now. The passage is interesting for the argument used by Robinson. He admits that clearly Paul understands this 'in terms of temporal brevity'. But then he goes on: 'What is thus contracted is not χρόνος, duration, but καιρός, the time for decision. In other words, the time for response and commitment is cut down.'[2] Now it is true that Paul might be said to have written the sentence in general because the time for decision is short or contracted, though this itself is a somewhat forced exegesis; the context rather suggests that the question whether to marry or not to marry is unimportant because of the shortness of the time available. But even if we admit that urgency of decision is a main motive in the uttering of the sentence in the first place, it is clear that this has nothing to do with the choice of καιρός as the word for 'time', and that the choice of καιρός rather than χρόνος does nothing at all to prove such a case. Robinson's argument involves the extension of the sense 'decisive time' for καιρός to that of 'time for decision', in itself a pretty doubtful shift, and one which is particularly improbable in the syntactical context of the verse. The point clearly is that the time from now until the end is a short one; and it is most probable that the use of χρόνος would not have altered the sense of the sentence in any way at all.

(c) One may also mention the evidence of compounds like πρόσκαιρος. Although this adjective was current in contemporary usage (in Plutarch, for example) with the sense of roughly 'opportune', a sense agreeing with the common classical sense of καιρός, all cases in the Bible (3 in 4 Macc. and 4 in the NT) have the meaning 'temporary, lasting only a short time', which depends on the sense of καιρός as 'time' or perhaps 'moment'. In Luke 8.13, where the parallels in Matthew and Mark have πρόσκαιρος, the text reads οἳ πρὸς καιρὸν πιστεύουσιν, and clearly means 'who believe for the moment, for a short time'; it has

---

[1] Robinson's exegesis of καιροὶ ἴδιοι (op. cit., p. 46) as 'God's own to dispose' is an impossible one in every way.  [2] Op. cit., p. 52.

nothing to do with 'at the right time' or 'at the time of decision'. On the other hand εὔκαιρος means 'opportune'.

(d) We may add the use of καιρός for a season of the year like spring or summer, a sense which is highly probable at a number of places in the LXX and appears at Acts 14.17 and I Clem. 20.9. The καιρός here is a recurrent period.[1]

This is by no means all the evidence which bears on the sense of καιρός and χρόνος in the Greek Bible, and further considerations will be added later. But the main point has been abundantly established, namely that the correlation of two great conceptions of time with the two Greek words is thoroughly erroneous, and that all arguments about 'time' in biblical thought are misleading in such proportion as they depend upon this correlation.

As already stated, it is not our intention to discuss the work of Marsh and Robinson in general, although some comments on other aspects of their positions will be added later. Their work as a whole contains many valuable observations which are not affected by our criticisms here, and may well in general represent a perfectly reasonable and tenable philosophical-theological position. But it remains clear that in their work as it now stands the καιρός-χρόνος distinction as presented by them is fundamental. It forms for them the means by which certain of their arguments are at crucial points attached to the biblical detail, and they return to it again and again.

*Note on the history of the καιρός-χρόνος distinction in the texts where the words are juxtaposed (Acts 1.7; I Thess. 5.1)*

In the post-Reformation period we find among some considerable authors an interpretation by which καιρός differs from χρόνος not by being a 'critical' or 'opportune' time but by being a *shorter* time. Thus Grotius says, on Acts 1.7[2]: 'χρόνοι sunt maiora temporum spatia, ut anni; καιροί minora, ut menses et dies. habes sic distincta Dan. 7.12 in Graeco'. We note that Grotius based his interpretation on actual biblical usage, whether or not we agree with his interpretation wholly. Bengel follows Grotius

---

[1] Cases like this and the preceding, along with certain other usages, may have been influential in producing the idea, which we shall notice later, that in the Bible καιρός is distinct from χρόνος not by being 'critical' but by being a *shorter* time.

[2] H. Grotius, *Annotationes in novum testamentum* (Groningen, 1829), vii. 171.

in his usual succinct manner: 'χρόνων partes καιροί'. This interpretation, though it may have gone too far in insisting on the
'short time' sense of καιρός, was certainly right in avoiding
domination by the 'critical time' and 'opportunity' sense.

An unfortunate reaction takes place with Archbishop Trench,[1]
who declares that 'this distinction, if not inaccurate, is certainly
insufficient, and altogether fails to reach the heart of the matter'.
He goes over to the 'critical time' sense for καιρός, maintaining
that this is the 'real' meaning. This means in effect, of course,
merely an adherence to the sense in classical Greek. Trench also
went on to talk of great critical moments of history, such as the
conversion of the Roman Empire to Christianity, or the Re-
formation, as examples of the sort of thing that would be meant.
Trench's position was followed, but somewhat modified, by
Lightfoot,[2] who says that the 'real difference' was that καιρός
indicated a quality and χρόνος a quantity, as stated by Ammonius.

Both Trench and Lightfoot, apart from appealing to classical
sources, cite a passage of St Augustine in which he did the same
thing. Commenting on the collocation of the two words at
Acts 1.6 f., Augustine says: 'nostri autem utrumque hoc verbum
tempora appellant, sive χρόνους, sive καιρούς, cum habeant haec
duo inter se non negligendam differentiam. καιρούς quippe
appellant Graeci tempora quaedam, non tamen quae in spatiorum
voluminibus transeunt, sed quae in rebus ad aliquid opportunis
vel importunis sentiuntur . . . χρόνους autem ipsa spatia temporum
vocant.'[3] This is just the view which Trench and Lightfoot
make their own. But apart from the fact that Augustine is appealing not to biblical but to classical Greek usage, one must remember that his argument, here quoted, is directed against a
special problem. There was a certain group of chiliastically-
minded Christians who interpreted the words of Matt. 24.36 in
such a way 'ut putent se posse computare tempora; diem vero
tantummodo ipsum et horam neminem scire'. Presumably they
reckoned the long-term periods until the end of the world, the
'tempora', but admitted that they could not predict the exact
details. Against this position Augustine deploys Acts 1.6 f., so

[1] R. C. Trench, *Synonyms of the New Testament* (London, 1865), pp. 200-3.
[2] J. B. Lightfoot, *Notes on Epistles of St Paul* (London, 1895) p. 70 f.
[3] PL xxxiii. 899 f. (*Ep.* 197. 2).

as to make clear that the 'times' there include both the long-term periods and the elements of opportunity which affect the choice of the moment. But the special purpose of this argument diminishes its evidential value for our purpose. We shall see later that St Jerome, writing not in the context of such special argument but in that of steady biblical translation and commentary, handles καιρός in a quite different way.

Basically both Trench and Lightfoot try to state the meaning of the words, and the difference between them, as if no difference was made by the period of usage or by the syntactical context.

In commentary literature on the two passages, Lightfoot was followed in a somewhat modified form by Milligan (1908), p. 63; and in more recent times, partly under the influence of Cullmann's theories, something of the kind seems to be hesitantly favoured by the careful Roman Catholic work of Rigaux (1956), p. 553 f.

But there has also been a line of interpreters who maintained either that the two words had much the same meaning, or at any rate that the difference was not that indicated by the 'critical time' sense of καιρός. We have already seen that this line included Blass and Jackson and Lake. Further, E. von Dobschütz (1909), p. 204, called the construction a pleonastic one, or a hendiadys, and said that the words did not indicate different objects. J. E. Frame in the ICC commentary (1912), p. 180, stated that 'in Jewish usage the terms are interchangeable'. In more recent times the same position is taken by C. Masson (1957), p. 66.

Although this latter point of view has not always been quite correctly phrased, there can be no doubt that, placed in contrast with the other, it shows an absolute superiority, working as it does from actual biblical usage, while the other depends on a 'real' meaning (based in fact on classical usage) and on the combination of this 'real meaning' with general considerations of biblical thought.

The correct understanding of the passages is well shown by the rendering of the NEB with 'dates and times'.

# III

## THE STRUCTURE OF *ΚΑΙΡΟΣ* AND *ΑΙΩΝ* —CULLMANN AND OTHERS

In his famous book on the biblical understanding of time and its relation to Christology[1] Cullmann also begins from a study of the temporal terms used in the New Testament, and like Marsh and Robinson he makes his main point through a comparison of two words, the relation of which forms or represents in his opinion the basic structure of New Testament thought about time. For Cullmann, however, the two basic words are καιρός and αἰών. Although at this point he lists a number of other words, such as ἡμέρα, ὥρα, and χρόνος, and maintains that the frequent appearance in the NT of all available words for time is characteristic of the keen interest of the NT in that subject, he says that 'the two concepts which most clearly illustrate the New Testament view of time are those usually represented by καιρός and αἰών.' Then he admits that these terms 'can also be used in the New Testament without any special theological reference'. Then follows the great distinction:

'The characteristic thing about καιρός is that it refers to a *point of time* defined by its content, while αἰών designates a *duration of time*, a limited or unlimited *extent of time*. In the New Testament both words serve, in a manner which corresponds remarkably well to the reality, for the characterization of that time which is filled out by the salvation history.'[2]

Less time need be spent on the discussion of this distinction itself than on that between καιρός and χρόνος, partly because some of the arguments used to show the weakness and inaccuracy of that distinction have the same effect on this one. This one is on the whole rather less erroneous than the former, for it is at least true that αἰών never means a point of time, even if it is not true that καιρός never means a period or extent of time.

---

[1] *Christus und die Zeit* (Zollikon, 1946); E.T., *Christ and Time* (London, 1951). I have myself translated anew some of the passages quoted, not because I wish to criticize Filson's translation, but in order to secure uniformity with my words elsewhere; for example, I use 'concept' for *Begriff* uniformly, as in my *Semantics*, p. 210 f. References are given for both editions.

[2] Op. cit., p. 33; E.T., p. 39.

We have already quoted two passages in the NT where καιρός meaning 'time' is used of an age or period, namely Heb. 9.9 παραβολὴ εἰς τὸν καιρὸν τὸν ἐνεστηκότα 'a similitude for the present time', and Mark 10.30 (Luke 18.30), νῦν ἐν τῷ καιρῷ τούτῳ 'now in this time' contrasted with ἐν τῷ αἰῶνι τῷ ἐρχομένῳ 'in the coming age'. In these passages very little appreciable difference would be made by substituting αἰών for καιρός. For the phrase of Heb. 9.9 cf. the ἐκ τοῦ αἰῶνος τοῦ ἐνεστῶτος πονηροῦ of Gal. 1.4, unless this should be translated as 'world'; in any case the participle clearly indicates something present for some continuous duration. If we include LXX evidence, I have already shown that numerous cases exist where καιρός means 'time' in the sense of a 'period' of some kind, and cases are especially numerous in Daniel, a book which was of course of special interest to the early Christians because of its apocalyptic emphasis. The common NT phrase ἡ συντέλεια τοῦ αἰῶνος 'the end of the age' or 'the end of the world' can be paralleled by phrases like ἕως συντελείας καιροῦ 'until the end of a time' at Dan. 9.27 Theod. or by καὶ κατὰ συντέλειαν καιρῶν καὶ μετὰ ἑπτὰ καὶ ἑβδομήκοντα καιροὺς καὶ ἑξήκοντα δύο ἔτη 'and at the end of times and after 77 times and 62 years' at Dan. 9.27 LXX. I do not suggest that καιρός here could be replaced by αἰών with no change of sense, but think it impossible that readers of such passages could fail to understand that a καιρός like an αἰών was something which could have a duration and a 'finish'.

We conclude then that in biblical Greek usage it is impossible to state the distinction between καιρός and αἰών as that between 'moment' and 'age' or 'period'. On the other hand it is also fairly clear that, when an 'age' in the sense of one of the great epochs of the world's duration is meant, it is much more *common* to use αἰών than to use καιρός.

Cullmann's statement of a fundamental distinction between καιρός and αἰών, a distinction which, we must remember, is the basis of Cullmann's lexical structure and is supposed to bring out the essence of the New Testament understanding of time, thus fails to fit the facts. There are however one or two other points in Cullmann's statements which may be construed as implicitly admitting that the facts do not entirely fit his fundamental scheme. If so, this might mean that Cullmann recognized the existence of anomalous cases, but thought that his theory was correct in essence. We may now examine these two qualifications, which in any case present points of general interest and importance.

The first is Cullmann's statement, already registered above, that his words καιρός and αἰών also occur without any special theological reference. What is the importance of this statement? It would be possible, though hardly probable, that it should be construed to mean that the opposition stated to exist between καιρός and αἰών, the opposition which is the backbone of Cullmann's lexical structure, exists only where a special theological emphasis is intended. To this possible interpretation there are, however, two immediate objections:

(a) The examples, which I have cited above in order to show that Cullmann's basic distinction does not fit the facts of usage, are themselves richly theological examples by any criterion.

(b) In any case it is impossible in this discussion to admit the validity of any exception on the grounds that the words are not used with a special theological emphasis. The whole case being argued is that the Bible has, and normally and constantly displays, a particular conception of time, which can be traced in its lexical stock and which forms an essential background or presupposition for the understanding of its theology. It is therefore naturally impossible to except any example of usage from full consideration on the grounds that it is 'merely temporal' and therefore not of theological significance. This point is of importance, whether or not we regard it as the meaning of Cullmann's sentence at this point.[1]

The second point made by Cullmann which may be perhaps construed as a qualification by him of his own basic distinction between καιρός and αἰών is his mention of 'the two concepts' which are 'usually' represented by these Greek words. In the

---

[1] Whether or not this argument is intended or implied by Cullmann at this point, it is one which has in fact been used. It appears for example in Delling's *TWNT* article on καιρός (*TWNT* iii. 460). Delling's favourite sense for this word is that of 'decisive moment'. In the LXX section of the article, as elsewhere, he emphasizes this sense. But he realizes that there are very many cases where this sense is impossible. He says therefore that 'the purely temporal sense is much more numerous, but has little theological interest', and goes on to register it in small print. This enables him to pick out the sense of 'decisive moment' as the essential contribution of the LXX to the theological understanding of time. For the NT he follows essentially the same procedure. See *Semantics*, p. 225 f.

writer's submission, it is perhaps possible, but certainly far from probable, that Cullmann means that the 'concept' which is usually represented by αἰών may occasionally be represented by καιρός. There is no sign in the following argument that this possibility is envisaged, and no sign that for any case of καιρός Cullmann considers possible the sense of 'time' generally or of 'period of time', the sense, that is, which he regards as the opposite of the sense of καιρός. In all probability the exceptions envisaged by the 'usually' of Cullmann are not cases where καιρός means 'time' or 'period of time' rather than 'moment of time' or 'decisive moment', but cases where the so-called 'kairos concept' or 'the conception of special moments of time' is expressed by words other than καιρός, such as ὥρα, ἡμέρα or even χρόνος.[1] In other words, the exceptions implied are not exceptions to the mutually exclusive distinction of καιρός and αἰών, but cases where one of these words is replaced by yet another word; a replacement which, as Cullmann sees it, does nothing to affect the validity of his great two-dimensional structure.

We may conclude then that Cullmann's treatment does not envisage any real exception to his basic distinction, namely that καιρός 'refers to a *point of time* defined by its content, while αἰών designates a *duration of time*'. But the fact that this basic distinction is not a correct one severely damages Cullmann's whole case. Cullmann's entire chapter on the words for time is intended to set forth 'the meaning of the New Testament terminology for time', and the whole point is to show that 'this terminology itself expresses the distinctive quality of primitive Christian thought about time'.[2] The distinction between καιρός and αἰών is the basic element in this lexical structure, and is claimed to 'illustrate most clearly the New Testament conception of time'. The fact that significant cases occur where καιρός does not mean 'a point of time defined by its content', and that the opposition to αἰών is thus quite wrongly stated by Cullmann, thus greatly weakens or entirely destroys the connection between his theory of time and the facts of lexical usage in the New Testament.

We shall now make an investigation of two procedures, the

---

[1] See Cullmann, pp. 36 f. and 42; E.T., pp. 43 f. and 49.
[2] Cullmann, p. 32; E.T., p. 39.

use of which in Cullmann's presentation has the effect of concealing the weakness of his theory here. They are: (a) what I
shall call 'the concept method', and (b) the use of transliterated
Greek words along with the avoidance of translation. For each
of these we shall go beyond Cullmann's own argument to say
something of the use of these procedures elsewhere.

I have already elsewhere called attention to the great prevalence, in theological studies of linguistic data, of the habit of
saying 'concept' for the reality we would normally call a 'word'.[1]
It is not necessary here to insist that it must be *wrong* to talk about
'the Hebrew concept *ga'al*' or 'the Greek concept μάταιος'. Let
it suffice that it is certainly wrong to speak thus if it leads to, or
permits, or conceals, any mishandling of linguistic facts. That
this is in fact the case with καιρός is in my opinion demonstrable.
But in general it will surely be desirable, in order to avoid confusion, that we should not say 'concept' where we mean 'word',
and that where we do say 'concept' we should understand that
we mean something other than a word.

The general confusions and dangers of what I shall henceforth
for brevity call 'the concept method' need not be discussed here.
There are, however, certain ways in which its weaknesses are
especially well demonstrated in the treatment of καιρός.

Firstly, the concept method breaks down wherever it meets
with polysemy, that is, with the fact of one word which has two
or more different senses. Surely one could not deal with modern
English 'nice' by putting its main senses ('pleasant, attractive'
and 'precise, exact') under the heading of 'the English nice-
concept'. Such a course would seem artificial in the extreme.
Nevertheless the same procedure is not uncommon in modern
theological treatments of biblical language. We have already
seen from our discussion of καιρός and χρόνος in the New Testament that καιρός sometimes means 'right or decisive time' and
sometimes 'time', the reference of the latter being to a period,
without any element of decisiveness. We cannot legitimately
subsume these under a single 'kairos concept'. If on the other
hand we simply proceed to talk about 'the kairos concept', and
ignore the well-known fact of the polysemy of the word, then
we shall certainly produce the effect, intentionally or unintention-

---

[1] *Semantics*, p. 209 ff.

ally, that our readers will not notice that the general sense of 'time' exists for this word; they will assume that 'the kairos concept' is identical with that sense of 'decisive time', which would not appear (say) in 'the chronos concept', and then will assume that the appearance of the word καιρός is naturally an occurrence of 'the kairos concept' and a reference therefore to some kind of 'critical' or 'decisive' time. Or, to put the matter conversely, if we feel we must go beyond speaking about 'mere' words, and must therefore speak about the concepts represented by them, then we must say that New Testament καιρός represents more than one concept, and at a minimum two, namely that of 'right time' and that of 'time' in general—the latter a concept commonly indicated also by χρόνος. But in the concept method as used by Cullmann and others, and as criticized here, the mention of the 'kairos concept', along with the knowledge that in the New Testament at some places, and in earlier Greek in most or all places, καιρός in a temporal sense is distinctive from χρόνος by having the sense of 'decisive time', leads to the conclusion that (a) this is the normal sense in the New Testament; (b) cases where the 'non-decisive' sense might exist may legitimately be argued over into some kind of 'decisive time'; (c) this somehow fits into the general interest in decisive temporal events in Jewish-Christian theology. Thus there can be no doubt that when Cullmann speaks without embarrassment of 'this central New Testament temporal concept of the kairos'[1] he means the idea of 'a point of time defined by its content' as already stated, and leaves no room for cases where καιρός does not mean this.

There are, however, perhaps certain reasons why the weaknesses of the concept method have not been so immediately obvious to students as they would be, for example, in the case of the English word 'nice'. Firstly, as a matter of linguistic knowledge itself, the very strong evidence of the Hellenistic usage of καιρός might not be effective, either because of sheer ignorance of it, or because of a prejudice that Hellenistic usage might be quite irrelevant to the New Testament because the thought of the latter is so different from Hellenic thought. Secondly, as already stated, the known interest in decisive

[1] Cullmann, p. 33; E.T., p. 40.

historical events in the biblical tradition might incline people in favour of the distinctive meaning of 'decisive time'. Thirdly, it is well known that influential philosophic theologians like Tillich have long used the word 'kairos' when writing in English, with a sense something like 'moment of crisis', and it is likely that familiarity with this modern usage will have led some readers to expect this as the natural meaning of the ancient word. We must naturally grant philosophers like Tillich the right to use such a word if they feel it necessary, provided they make clear how they use it; but their usage of it is of course of no significance for the understanding of ancient texts.[1] Fourthly, the general vagueness and intangibility of the subject of time may have contributed to the failure to see the weakness of the concept method here.

More important than any of these, however, is another reason why the concept method is so important here. It is, in the writer's submission, a corollary of the attempt to clarify a subject by the building of a lexical structure, that the elements of that structure, namely the words, should be regarded as 'concepts'. The lexical structure of words must coincide with a mental structure of concepts, which latter is basic to biblical thought about time. It not only is natural, therefore, but from the dictates of the plan *must be* natural, that there should be a 'kairos concept'. It *must* be possible to conceptualize the words. If the possibility were envisaged that καιρός might sometimes mean Marsh's 'chronological time' and only *sometimes* mean his 'realistic time', or that it sometimes might mean a period and only sometimes Cullmann's 'point of time defined by its content', it would at once become impossible to connect with specific biblical words the conceptual contrast basic to the picture of time in both writers. A reliance on the concept method is a corollary of the general approach through a lexical structure. It may be suggested here that there are dangers in the concept method even where polysemy does not wholly damage it; but this need not be discussed here, since the polysemy of καιρός makes the concept method impossible in this case, and the phrase 'the kairos concept' has no intelligible meaning.

[1] It is unknown to the writer whether Tillich thought his use of 'kairos' to be in accordance with biblical usage or not.

Obviously, we should now consider whether Cullmann in his exposition has committed one of the faults to which we have called attention, namely that of pressing the sense of 'critical moment' or 'decisive moment' upon examples of καιρός in cases where this sense is uncertain, or even improbable or impossible. There is however a reason for leaving Cullmann's exposition for a moment, since we shall return to it in the treatment of the use of transliterated Greek words. Meanwhile we may perhaps point out a few cases where writers other than Cullmann have in fact pressed the sense of 'critical moment' or 'decisive moment' upon καιρός without good reason, or where other effects of the concept method can be seen in the discussion of this word. In particular a number of possible pitfalls can be seen in the work of Delling.

We may begin from the example which I have already criticized elsewhere,[1] namely Delling's treatment of Luke 4.13, where the devil departs from Jesus ἄχρι καιροῦ. This surely means 'for a (short) time', and is so rendered by competent dictionaries and translations. Bauer for example says 'until (another) time, for a while'.[2] But Delling interprets it as 'until the moment to be decided by God'[3]; cf. also Karl Barth, who says 'until the decisive moment'.[4] The sense which I have given is surely supported by Acts 13.11, where Elymas is to be τυφλός, μὴ βλέπων τὸν ἥλιον ἄχρι καιροῦ 'blind, not seeing the sun for some time', a case where the sense 'until the decisive moment' is surely very unlikely.[5] One may also cite:

Sir. 1.23-24 ἕως καιροῦ ἀνθέξεται ὁ μακρόθυμος
and ἕως καιροῦ κρύψει τοὺς λόγους αὐτοῦ.

Here the sense is more naturally understood as 'for a time, for the time being' than as 'until an opportunity comes' or 'until the decisive moment'. Similarly, at Tobit 14.4 (BA text)

ἐν δὲ τῇ Μηδίᾳ ἔσται εἰρήνη μᾶλλον ἕως καιροῦ

---

[1] *Semantics*, p. 225 f.
[2] Bauer, p. 395; unfortunately, in the article ἄχρι, p. 128, he gives the other interpretation, 'until a favourable time'.
[3] *TWNT* iii. 463.
[4] *KD* iv/1. 290; E.T., p. 264.
[5] RSV correctly says 'for a time' here, although at Luke 4.13 it offers 'until an opportune time' for the same phrase, against AV and RV.

surely certainly has the meaning 'for some time', and the ἔως καιροῦ here probably does not greatly differ in sense from the μέχρι χρόνου later in the same verse where we are told that the Jerusalem temple ἔρημος ἔσται μέχρι χρόνου.

In face of this evidence, then, it is probable that Delling has simply found the general idea of 'decisive moment' in connection with the Gospel events so attractive that he could not resist pressing Luke 4.13 into it, although the evidence for the general and non-decisive sense of καιρός was well known to him and registered in his *TWNT* article.

It may be added that the preference for the sense of 'decisive time' may be distorting not only to cases of καιρός with the sense of 'time' in general but also to cases where the sense is the old and well-established one of 'opportunity'; for even 'opportunity' will not always fit into the kind of 'decisive time' which is the modern theological 'kairos concept'. Thus at Gal. 6.10 we have ὡς καιρὸν ἔχομεν, and the sense is 'as we have opportunity' (so AV, RV, RSV). Delling's comment[1] is: 'The Christian is in possession of this καιρός, that is, as pneumatic man he has the capability of recognizing it and realizing its (his?) will.' This is surely a very exaggerated force to give to καιρὸν ἔχειν; it is produced by the determination of the noun as '*this* καιρός', by the ignoring of the special syntactic setting of the phrase with ἔχειν, and by the domination of the whole picture by the supposed 'New Testament kairos concept'. It is a great inflation of the phrase 'as we have opportunity' to produce from it the picture of the Christian who 'possesses the kairos'.

We see then that the concept method may have the effect of forcing into a particular sense, i.e. the sense supposed to inhere in the 'concept', examples of a word where a different sense is likely or even certain. There are, however, other ways in which the concept method can damage the assessment of linguistic evidence. These can also be exemplified from Delling. Three main points may be noticed.

Firstly, the interpreter, armed with his understanding of the 'kairos concept', may from time to time give the writers a lesson

---

[1] *TWNT* iii. 461; cf. *Zeitverständnis*, p. 88: 'The kairos provides the basis for the life in faith of the Christian, just as his moral action also flows from it; it has become his inner possession' (quoting Rom. 13.11 and Gal. 6.10).

in how they should have used their language. Thus in the Lucan writings, according to Delling, the statistics of καιρός and χρόνος are:

| Gospel | καιρός | 13 | χρόνος | 7 |
| Acts | καιρός | 9 | χρόνος | 17. |

The greater frequency of καιρός in the Gospel, Delling holds, is because Luke found it in his sources and therefore used it again; but he himself liked χρόνος, and in the Acts used it frequently. Thus, Delling maintains, Luke 'fails to take the opportunity— quite unambiguously offered by the reality described—to designate the events of the missionary history as καιροί, as fulfilled times of God'.[1]

Surely nothing can be more ill-conceived than thus to produce a theological 'kairos concept', and then to use this as a standard to measure how far a writer has really grasped the centre of New Testament thinking; and in this case to judge that Luke was naïve and unreflecting in his understanding of time, and failed to use the opportunities given to him to express the characteristic aspects of time in the New Testament.

This is by no means the only case where Delling uses such reasoning. Discussing the Old Testament phrase 'the Day of the Lord', he similarly criticizes the LXX for 'thoughtlessly' translating with ἡμέρα θεοῦ or ἡμέρα κυρίου while 'a real sympathy with the Greek linguistic sense would have made καιρός natural here'.[2]

Against this we need have no hesitation in affirming that both the LXX and St Luke knew what they were doing. The procedures which Delling suggests they might have taken are in fact fictions produced from the prominence of the 'kairos concept' in Delling's mind, and there is no evidence from usage that the ideas involved could on the basis of current usage of καιρός have been handled with this word at all.

Secondly, we see in Delling a depreciation of actual linguistic usage as 'naïve' or 'unreflecting' where the use of καιρός does not coincide with the 'kairos concept' supposedly represented by it. Thus he notes correctly the fact, which we have already outlined, that in Hellenistic and New Testament Greek καιρός frequently means not a critical moment but a period of greater or less duration. But in his judgement this common usage is 'a naïve way of

[1] *Zeitverständnis*, p. 80.       [2] *Zeitverständnis*, p. 85.

using the temporal concepts, one which unreflectingly pre-supposes time as something naturally given, and which cannot make an abstraction from it'. Thus this naïve usage has the serious effect of obscuring the distinctiveness of the Jewish thought structure from the Greek.[1] This is the logic by which it becomes natural for Delling to write his *TWNT* article around the sense of 'decisive moment', giving no important place to the sense of 'period' and tucking it away in small print. The giving of real attention to this 'naïve' usage would only obscure the depths of biblical thought about time.

Thirdly, the concept method may lead to an account of words which attributes to them potential meanings, or actual meanings other than those in the minds of the persons who used them. Thus Delling thinks that καιρός had a potentiality which the Greeks, for reasons inherent in the structure of their thought, had never perceived, and which was seen and brought out by the New Testament writers: 'The New Testament writers felt strongly the ethical and religious possibilities for development in καιρός, which in the development within Greek had not received their full force or recognition (*die rechte Geltung*).'[2] Surely nothing could be more artificial than a theory according to which the full potentiality of καιρός in a sense of 'decisive moment' is neglected by the classical Greeks and first developed only by the New Testament, when in fact it is in classical Greek that this sense is predominant, and when in New Testament usage it continues to appear only in certain contexts.

Some of these judgements of Delling's are very much his own property, and it is not suggested here that they have had a general acceptance; certainly few have wanted to go so far as to teach the biblical writers how they should have expressed themselves. It remains true however that Delling's *Zeitverständnis* is not by any means a bad book; it continues to be quoted, and might well have had much greater influence outside of Germany but for its being published during the war; and, as has been shown, its methods are basic also to Delling's *TWNT* article on καιρός. It shows a much greater understanding of the opinions held of time in Greek philosophy than is shown by Cullmann himself. But our main point is that Delling's procedures either

[1] Ibid., p. 78.   [2] *Zeitverständnis*, p. 84.

arise from, or depend for their continued existence and non-correction upon, the concept method of handling lexical items. This method is very common in theological treatments of language, and appears never to have had either proper justification or proper criticism.

It remains to mention one other danger which may arise from the concept method, namely the possibility that we shall proceed to speak of the words, or of the word concepts like 'the kairos concept', as if they were realities of some kind, and not only functional units used in such and such a way in the linguistic system of such a time. This procedure may best be called hypostatization, and the writer has elsewhere published a discussion of it, in particular in its reference to theological treatments of language.[1] Suffice it to give an example from the discussion of time: according to Robinson, the New Testament writers may sometimes have 'written as though it were chronos that determined kairos and not vice versa'.[2] This is surely speaking as if these two words were realities of some kind which might interact in some way on one another. Other examples of this, and further discussion of the problems which arise from it, will appear later.

We come now to the second procedure used by Cullmann which we have undertaken to criticize, namely the use of transliterated Greek words along with the avoidance of translation. This particularly irritating procedure appears throughout Cullmann's discussion of the usage of καιρός, and adds to the difficulty of knowing whether Cullmann has pressed into the mould of the 'kairos concept' examples of καιρός where the sense of 'moment' or of 'decisive moment' is unlikely or disputed.

Thus in Cullmann's discussion a part of the important sentence at Acts 1.7 appears as 'the kairoi which the Father has fixed in his omnipotence'.[3] Similarly the καιροῖς ἰδίοις of I Tim. 6.15 appears as 'at appropriate kairoi' (*zu geeigneten Kairoi*).[4] This treatment by a kind of semi-translation is given in fact to all the New Testament sentences treated in the section on καιρός.

Now in the context, where Cullmann has just stated that the

---

[1] See the writer's 'Hypostatization of Linguistic Phenomena in Modern Theological Interpretation', *JSS*, forthcoming.　　[2] Robinson, op. cit., p. 53.
[3] Cullmann, p. 34; E.T., p. 40.　　[4] Ibid., p. 34; E.T., p. 40.

characteristic thing about the kairos is that it is a point of time defined by its content, the reader will almost certainly understand that this is the meaning of καιρός in the sentences where Cullmann thus transliterates it. But Cullmann in such cases does not at all discuss the arguments for and against this sense. We have already seen that at Acts 1.7, for example, this sense has long been doubted or rejected by many competent scholars. Their opinions are in fact completely ignored by Cullmann. By not translating the Greek word, and by adopting 'kairos' as a German or an English word into his version of the sentence, he has quite evaded the question of its sense here; but at the same time he has directed the attention of the unwary reader to the 'kairos concept' just set up, and has seemed to be giving supporting evidence for the importance of this concept. Meanwhile also he has created a quite unjustified impression of exact and accurate argument—for who will deny that 'kairos' in Cullmann's German or English is represented by καιρός in the Greek text? Moreover, he has forced the passage into his case, as also with I Thess. 5.1, by simply omitting the χρόνοι which occurs along with the καιροί in both. How could he do justice to its presence when he has just maintained that 'it is not all the fragments of ongoing time that constitute salvation history in the narrower sense, but rather these specific points, these kairoi, singled out from time as a whole'?[1]

We thus see that Cullmann by his device of transliteration by-passes the difficult, uncertain and often tortuous arguments which are required to establish the sense of 'decisive moment' in such passages—arguments the very presentation of which could not but call attention to the strength of the case in the opposite direction—and rather tacitly produces an impression that 'the New Testament kairos concept' must be meant.

Apart, however, from the ways in which this sort of semi-translation is used in this part of Cullmann's argument, there are considerable dangers in its use in general. One of these dangers is that a word which itself is undoubtedly biblical may, by being planted in an English sentence, be placed in a new syntactical context, in which the new environment shifts the meaning of the word to one which it never has in its original biblical syntactical

[1] Cullmann, p. 33; E.T., p. 40.

context. Thus, to take an example, there is no harm in paraphrasing Ex. 33.18 into: 'Moses asked that he might see the doxa of God.' But there is very considerable harm in producing sentences such as: 'Israel is called to be laos but desires to be ethnos'. The former is a valid reproduction in English of a syntactical context in which δόξα stood in Greek, and one therefore which will not seriously misrepresent the sense of the original, while giving us useful information about the Greek word used therein. The latter is a purely fictitious context, in which the biblical words are used, but used in relations (especially in consequence of their indetermination here) which give a quite false impression of their meaning in actual contexts.[1]

Returning to Cullmann, we see that similar faults appear in his presentation of the words for time. Thus in discussing I Peter 1.11 he says: 'So also the First Epistle of Peter speaks of the time of the Christ-events, already complete from the author's point of view, as a kairos. The prophets of the Old Testament "searched to learn what or what kind of kairos was meant by the Spirit within them, when it testified to the sufferings destined for Christ and to the subsequent glory".'[2]

Now the basic proposition of this argument could be set out thus:

'Peter said that the time of Christ was a kairos.' But we see, as soon as it is formulated, how far this is from the kind of syntactical context in which καιρός is actually used. The more general sense of 'moment' or 'time' is not intelligible in Cullmann's sentence, and there is no doubt that he intends the sense of 'decisive time'. But the sense of 'decisive time' or 'right time' is surely definite; i.e. it is used for '*the* right time' for something, and not '*a* right time' or '*a* decisive time'. If Cullmann means that it is 'a kairos' in the sense of 'one of the kairoi or decisive turning-points of the salvation history' (and this is the probable interpretation of his argument), then one must simply say that this picture of a history composed of a series of kairoi is produced

---

[1] For the case of *laos* and *ethnos*, see *Semantics*, p. 234 f. For criticism in general of the use of transliteration in similar situations, see Bruno Snell in *GGA* cxcvii (1935) 331 f., and A. D. Nock in *Gnomon* xxvii (1955) 567. Snell says that 'if we describe Greek things by scattering in a few Greek words, we speak a pseudo-German and a pseudo-Greek, a language which does not permit precise thought'. How true this is of much modern theology!

[2] Cullmann, p. 34 f.; E.T., p. 41.

by stringing together a number of different passages, each of which contains the word καιρός with some reference to a theologically important event, and does not rest on any actual usage of καιροί in the plural meaning 'decisive moments in the salvation history'. In view of the numerous references in modern theological literature[1] to a 'series of kairoi' and the like, it cannot be sufficiently emphasized that this phrase produces a sense, and a kind of syntactical context, in which the biblical word καιρός does not occur at all. Such a statement as

'The story of the people of God is full of crises, καιροί, "decisive moments" '[2]

makes no contact with the actual usage of the word καιρός in the biblical literature at all, and would be much less misleading for biblical interpretation if the Greek word were omitted. There are no contexts in biblical Greek in which a 'series of καιροί' appears in usage, just as there is no context in which anyone says that 'this time is a kairos' in the sense that 'this time is a critical one'.

The misleading effect of transliteration appears also in Cullmann's later generalization[3] that 'in the past, the present and the future there are special divine kairoi, by the joining of which the redemptive line arises'. It is surely a most precarious exegesis of I Tim. 2.6, Χριστὸς Ἰησοῦς, ὁ δοὺς ἑαυτὸν ἀντίλυτρον ὑπὲρ πάντων, τὸ μαρτύριον καιροῖς ἰδίοις, to say that this clearly shows us 'the necessity to connect the kairoi with one another'. This is especially so if Filson is right, as he may well be, in interpreting Cullmann's translation of the last four words as 'as a witness to appropriate kairoi still to come' (German text 'zum Zeugnis geeigneter Kairoi'). This interpretation in a future sense has surely

---

[1] See for example Eichrodt in *ThZ* xii (1956) 118, 121; Ratschow in *ZThK* li (1954) 381 f.; von Rad, *Theologie*, ii. 114; K. Löwith, *Meaning in History* (Chicago, 1949), p. 185 f. Minear in *Eyes of Faith*, p. 110 n., says: 'Because of the wide divergence between the biblical and non-biblical connotations of the term "time", it is necessary in the following discussion to use the Greek word *kairos* to indicate the biblical connotations of a purpose-created and purpose-charged time.' But while there may or may not be a case for such a distinctive term in a modern discussion, we must be clear that the function of the word καιρός in biblical texts is *not* that of providing such a distinctive term; and that indeed it is in *non-biblical* texts that its use as something entirely distinctive from χρόνος is much more clear than in biblical. Cf. also H. Diem, *Dogmatics* (Edinburgh, 1959), p. 135, with reference to Käsemann; J. Daniélou, *The Lord of History* (London, 1958), pp. 32 f., 198; B.S. Childs, *Myth and Reality in the Old Testament* (London, 1960), pp. 72-83.

[2] G. A. F. Knight, *A Christian Theology of the Old Testament* (London, 1959), p. 316.

[3] Cullmann, p. 36; E.T., p. 43.

very little support in its favour, and is implicitly contradicted by many works of reference.[1]

This passage, however, is of extreme importance in the construction of Cullmann's argument. This interpretation of the καιροῖς ἰδίοις of the Pastoral Epistles is the only linguistic evidence produced by Cullmann for the existence of a line which joins up the kairoi. But for this line there would only be separate kairoi. The line is fundamental for Cullmann's later argument.

We conclude, then, that Cullmann, though he does not *argue* cases of καιρός into the sense of 'decisive time' where another sense would be possible, tacitly produces this impression by his use of 'kairos' in semi-translation.

One other point in Cullmann's treatment of καιρός remains to be mentioned. He distinguishes between a secular usage and a *heilsgeschichtlich* one, one related to the salvation history. The former means the moment of time specially favourable for an undertaking, but purely human considerations have led to its being deemed thus favourable. Such an example is the καιρὸν δὲ μεταλαβὼν μετακαλέσομαί σε of Felix to Paul (Acts 24.25), 'I shall summon you again when I next have a suitable time'. The latter on the other hand is determined not by human considerations but by a divine decision 'which makes this or that date into a kairos, and that in reference to the execution of God's plan of salvation'.[2] God chooses the times which he will make into the kairoi of the salvation history, and from a human angle this choice is arbitrary.

Later this distinction is used in the interpretation of John 7.6, ὁ καιρὸς ὁ ἐμὸς οὔπω πάρεστιν, ὁ δὲ καιρὸς ὁ ὑμέτερος πάντοτέ ἐστιν ἕτοιμος. This is supposed by Cullmann to mean that for Jesus' brothers there is no kairos in the New Testament sense of one related to the salvation history, no moments specially fixed and prepared beforehand; they have only kairos in the secular sense, where man decides whether a kairos is favourable or not; thus

---

[1] Cullmann's exegesis of the plural καιροί in cases like I Tim. 2.6, as representing a *plurality* of divinely planned moments, is rightly questioned by Bultmann in *ThLZ* lxxiii (1948) 663. I have already pointed out that the plural in the general sense of 'time', and with no element either of decisiveness or of plurality in meaning, is well established in Hellenistic usage and in the LXX. For other interpretations of καιροῖς ἰδίοις in these passages compare Bauer, pp. 370, 495 and translations like AV, RV, RSV.

[2] Cullmann, p. 33; E.T., p. 39 f.

the brothers can go up to Jerusalem at any time, while Jesus must await the kairos fixed by God.[1]

Whatever we may think of this exegesis of an admittedly rather obscure passage, the basic distinction is probably a confusion between what is generally true or believed to be true about God in relation to time and what is meant by the word καιρός itself. While no one would doubt that the early Christians believed that the times of important events were chosen by God, it is surely quite artificial to suppose that the use of the word καιρός itself forces upon the hearer or reader an obvious choice between human considerations and divine agency in determining that such and such a moment is 'the right time'. In any case Cullmann's discussion of what 'makes a date into a kairos' is another example of a quite artificially invented syntactical context. We have plenty of cases where it is said that 'God appointed these times' or something like it, and where καιρός is suitably used, being in any case the commoner word for 'time' in such contexts, but sentences such as 'God made this day into a καιρός' or 'God chose this day to be one of his series of καιροί' simply do not occur; arguments derived from consideration of them are not arguments about New Testament language at all, and cannot be used legitimately for determining the meaning of καιρός in contexts where it does appear.

The main importance, then, of Cullmann's section on καιρός is his working out of the series of decisive moments or kairoi chosen by God, the joining together of which furnishes Cullmann with his line, so important for his understanding of time. We have seen, however, that his presentation of the series of kairoi is often far remote from actual usage of καιρός in the New Testament; we have also seen that the attempt to use the καιροῖς ἰδίοις of the Pastoral Epistles as an example of the necessity of joining the kairoi together (an essential for the formation of the line, and the only example given) is quite improbable.

We now turn to Cullmann's discussion of αἰών, from which he will draw one of the conclusions of the greatest importance for his thesis as a whole, namely his opinion that 'eternity' is not something different from time, but is the whole of time, or end-

[1] Cullmann, p. 35 f.; E.T., p. 42.

less or unlimited time. It is explicitly claimed by him that an examination of the linguistic usage of αἰών itself demonstrates this to be so.[1] The arguments presented are the following two:

(a) The same word αἰών is used both for a precisely limited duration of time (where we might translate 'age') and for an unlimited and incalculable duration, where we are accustomed to translate by 'eternity'. This double sense is in Cullmann's opinion a guide to the relation between time and eternity in the New Testament.

(b) The plural αἰῶνες is particularly preferred where the sense is 'eternity'. Granted that rhetorical tendencies have had some effect, nevertheless 'the fact that one can speak of eternity in the plural proves that it does not signify cessation of time or time-lessness'. Rather it means an endless continuance of time, beyond the understanding of man, or better, 'the linking of an unlimited series of limited world periods, whose succession only God is able to survey'.[2]

We may first dispose of the second argument, namely, that the use of the plural for 'eternity' proves that continuance, and not cessation, of time is intended. To this argument in general one must simply ask 'Why?'. By what logic does it follow? Presumably the reasoning is that the plurality of the word implies some sort of variety within the reality so designated, and that variety and change indicate something like 'time'. But such an argument is a purely theoretical one, and takes no account of the circumstances in which plural forms may in fact be used in language. Quite in general one could think of various examples in different languages where a plural form does not mean a plurality and variety of the objects designated by the singular (cf. Lat. *copia* and *copiae*, *gratia* and *gratiae*, *castrum* and *castra*), or where no singular exists (one cannot argue that *tenebrae* means a plurality or variety of the object designated by the non-existent *tenebra*). In Hebrew in fact, quite apart from the words for time in particular, plurals which have no singular, or which do not mean a plurality of some reality designated by the singular, are quite common. There is therefore no general reason why we should accept Cullmann's argument that the use of the plural

---

[1] Cullmann, p. 39; E.T., p. 45 f.　　　[2] Ibid., p. 39; E.T., p. 46.

demonstrates that timelessness or cessation of time is not intended.

The point may be put, however, more precisely and particularly, in the suggestion that the plural αἰῶνες means 'the linking of an unlimited series of limited world periods'. The same point has been made by Sasse in the *TWNT* article αἰών as follows[1]: 'The plural in any case presupposes the knowledge of a plurality of αἰῶνες, of ages or extents of time, the infinite series of which constitutes eternity.'

But the suggestion in this form suffers from even more serious weaknesses. The use of the plural in contexts like εἰς τοὺς αἰῶνας τῶν αἰώνων, 'for ever and ever' or 'unto eternity', may in all probability be traced to, or at any rate greatly influenced by, the Hebrew and Aramaic phenomenon. This is probably true for Jewish-Christian religious usage, even if it should be possible to adduce examples from pre-Jewish and pre-Christian Greek usage.[2] Now there can be no real doubt that the Hebrew plural *'olamim* (and similarly the Aramaic) has nothing to do with the sense of 'age' or 'period' for the singular *'olam*. This latter sense is of very late origin in Hebrew, and is probably not extant in the Old Testament, unless in a difficult passage in Qohelet. The plural on the other hand is well established in classical Hebrew and shows every sign of being ancient, although in later times it became more frequent. There is in fact every reason to follow the classic Hebrew grammar of Gesenius-Kautzsch[3] and the most recent scholarly investigation of the subject[4] in regarding *'olamim* as some kind of 'plural of extension', and as having therefore nothing to do with a plurality of an object denoted by the singular *'olam*. This is strongly supported, as Jenni rightly sees, by the existence of cases like *lᵉ-neṣaḥ nᵉṣaḥim* in Isa. 34.10, also meaning 'for ever'. Phrases with *neṣaḥ* such as *la-neṣaḥ* are in a few LXX passages rendered by Greek phrases with αἰών, such as εἰς τὸν αἰῶνα or ἕως αἰῶνος. Now *neṣaḥ* means 'ever-

---

[1] *TWNT* i. 199.

[2] None are given by Preisigke, MM or Bauer. If we are right in tracing the plural principally to Hebrew influence, then LS are wrong in suggesting that the plural sense of 'the ages, eternity' is a development from the sense of 'space of time clearly defined and marked out, epoch, age' which they exemplify from the New Testament. The example from Philodemus which they cite is perhaps hardly a parallel.

[3] GK §124 a-b.

[4] E. Jenni, 'Das Wort *'olam* im alten Testament', *ZATW* lxiv (1952) 244 ff.

lastingness, perpetuity' (BDB), and is not a single object of which there can be a plurality. We may assert with assurance that Cullmann's argument that the use of the plural demonstrates the existence of a series of world periods, or demonstrates that eternity is not an absence of time, is entirely without validity.[1]

In any case we may question another statement of Cullmann, namely that the plural αἰῶνες is specially popular where eternity is being spoken of. It is difficult to know here what is the criterion for knowing which passages concern 'eternity', but if we include those passages with *'olam* in Hebrew or αἰών in Greek which concern the being, presence and decisions of God and can be translated as 'for ever', 'unto eternity', 'from all eternity' and the like, then we can state the evidence in general somewhat as follows. The plural is quite infrequent in the Hebrew Old Testament, 12 cases in all.[2] In the Aramaic sections of the Old Testament it is distinctly more frequent, 9 cases in Daniel. The LXX follows roughly the proportions of the Hebrew, except for Tobit, which brings in a big accession of plurals, as do also the apocryphal parts of Daniel. In the New Testament, allowing a considerable margin for cases which one might or might not wish to include, the figures for the singular and the plural would not seem to be far from equal. What is more significant is that certain writers prefer one type to the other. Thus all cases in the Gospel and Epistles of John are in the singular, all in the Apocalypse are plural; so are all cases in Paul, for II Cor. 9.9 is a quotation from the LXX and I Cor. 8.13, οὐ μὴ φάγω κρέα εἰς τὸν αἰῶνα, is perhaps hardly a case of 'eternity'. It is perhaps more probable to suggest that the difference is a stylistic one, and that the doxological style, to which many cases in Paul and the Apocalypse belong (as do many of the passages of the Greek Daniel already

---

[1] Marsh (*Fulness*, p. 174 f.) rightly sees that there is a fault in Cullmann's argument from the plural; but his own argument, that the phenomenon is a poetical taking of liberties with language in order to express what transcends the temporal categories of human speech, is remote from a realistic explanation of the linguistic facts, and his suggestion that the plural shows 'that God's time is other than ours' is quite as improbable as Cullmann's. Marsh's argument is not an entirely new one; for an earlier version, see Orelli, p. 81.

[2] Jenni, pp. 222 f., 243. The relative infrequency of the plural in Hebrew is surely an argument against Marsh's idea that its origin lies in the attempt to express the transcendent; for the problems of doing so must have been very much more common than the very occasional passages where the plural is used. In any case the example Marsh gives, Ps. 90.2, does not in fact use the plural.

mentioned), prefers the plural. In any case we can hardly state generally that the plural is preferred for 'eternity'.[1]

We now turn to the criticism of Cullmann's other point, namely that the use of αἰών both for an 'age' or limited duration of time and for unlimited duration or 'eternity' proves that 'eternity' in the New Testament is not something different from time but is unlimited time. We may begin in a quite general way by pointing out that if a word has two senses, as is agreed to be the case here, this can by no means be taken to show without further ado that the objects meant by the word in these two senses are essentially akin. At the most it will mean that there is some point of contact or similarity; but this point may lie in the past relative to the period of usage actually under discussion. One would do well to be on one's guard especially in cases of words for intangibles like 'eternity', where there is always a chance, to say the least of it, that semantic changes have extended the senses of words from more tangible references (though one would be unwise to say that this *must* happen either). One would not suppose that, because 'matter' occurs in 'a matter for discussion' and in 'the universe is composed of matter', therefore the subject of discussion is, or is akin to, the constituent of the universe. All this is preliminary, and simply shows that this argument of Cullmann's cannot be allowed to pass without challenge and examination. Cullmann himself however does little to support his point, other than to reiterate that the word for 'eternity' is the temporal term αἰών, and to reassert that in the New Testament time and eternity are not opposed; but whether this latter assertion is in fact supported by the linguistic facts is the question at issue.

We can best approach Cullmann's reasoning by studying the following section. He has been arguing that the difference between 'time' and 'eternity' consists only in this, that the former is limited and the latter unlimited, and he goes on:

---

[1] That the use of the plural is stylistic, and has nothing to do with the difficulties of expression where eternity or the transcendent is being spoken of, can be shown also from the Qumran documents. With some few exceptions, some of which can be explained by variation in OT usage from which phrases are quoted, we find an overwhelming preference for the singular in 1QH, 1QSb and CD. 1QM is equally solid for the plural. 1QS begins by using the plural almost entirely, and then after some brief hesitation settles down to the singular for the latter part.

'When the Christian writers in their statements go beyond calculable time in a backward direction, they write ἐκ τοῦ αἰῶνος ("out of the age"), ἀπ' αἰῶνος ("from the age"), or even ἀπὸ τῶν αἰώνων ("from the ages"); when they do so in a forward direction, they write εἰς αἰῶνα ("into the age") or εἰς τοὺς αἰῶνας ("into the ages").'[1]

Now it is very probable that these phrases should be treated as bound phrases, and the defect of Cullmann's treatment is that it works as if the semantic value of αἰών within these bound contexts is the same as that which it has when used freely in other contexts. The relation of the free sense of αἰών to the sense of εἰς τὸν αἰῶνα is not the same as that of οἰκία to εἰς τὴν οἰκίαν. Furthermore, we shall see that Cullmann's treatment here implies relation not only to a free sense, but to that free sense which occurs within the Greek New Testament, namely the sense found in ὁ αἰὼν οὗτος 'this age', which is not in fact the free sense of αἰών which, if any, should be related to the phrases quoted. This free sense would be the general and indeterminate sense of 'long space of time' (LS), from which we have for example the Hesiodic ἀπ' αἰῶνος 'of old' and the Aeschylean δι' αἰῶνος 'perpetually'. The sense of *space of time* clearly defined and marked out, *epoch, age* (LS) is a well-known New Testament sense of αἰών, but is not one relevant for the phrases quoted by Cullmann; it is neither a sense intended by the users of these phrases, nor is it a sense from which these phrases have historically arisen. But such translations as 'from the age' and 'into the age' clearly depend upon this sense, and, as Cullmann shows by his further use of the phrases, his intention is to interpret them as something like 'from that precisely limited period which ended at the creation of the world' and 'into that precisely limited period which begins with the eschatological events'. But the sense of αἰών in the phrases is not that of 'a precisely limited period' at all. Cullmann's translations are not only literalizing translations which give a false picture of the sense,

---

[1] Cullmann, p. 40; E.T., p. 46. The translations in brackets are taken from the English version, but certainly correctly represent Cullmann's intention, as is shown by what he writes later. I have however amended Filson's rendering in other respects, to give what seems to me a more probable interpretation of the German. Cf. also Cullmann's own French text, *Christ et le temps* (Paris, 1947), p. 33.

but their literalness depends on the use of another sense of αἰών altogether. In fact, to take εἰς τὸν αἰῶνα as our example, a far more accurate and less misleading translation will be given by rendering 'for ever' in most contexts, with 'never' in negative contexts; since for the past we cannot say 'from ever' in English, we could perhaps say 'from all time', or indeed 'from all eternity'. Phrases like 'from eternity', 'unto eternity' or 'for ever' are certainly a more adequate rendering of the Greek phrases than are 'from the age' or 'into the age'. To summarize: it is only in other syntactic contexts, and not in the ones quoted, that αἰών means a particular limited duration or 'age'. These phrases are therefore no evidence for the case that eternity differs from time only in being the unlimited entirety of time. The meaning has to be taken for the phrase as a whole.

This is so whether we look at the Hebrew or at the Greek background of αἰών as used in the New Testament. We shall consider first of all some points in classical Hebrew usage, which may well be deemed the chief background for New Testament usage. Here I simply follow Jenni in taking the sense of *'olam* to be adequately represented for most cases by 'the remotest time'. This is a temporal sense of a kind, but of a kind which receives no treatment by Cullmann in his study of the problems of time and eternity. As already stated, the sense of a 'period' or 'age' probably does not occur for *'olam* in classical Hebrew.[1] Thus *me-'olam* means 'from or since the earliest time', as in Joel 2.2 *kamohu lo nihya min-ha-'olam*, 'there has been none like it since the remotest time, ever'; *l^e-'olam* means 'to the farthest time, for ever'.

[1] Jenni in *ZATW* lxiv (1952) 246 f. It should be noted that Jenni's investigations make it difficult to sustain the interpretation of *'olam* given by Pedersen, *Israel* i-ii. 491: 'All the generations are fused into a great whole, wherein experiences are condensed. This concentrated time, into which all generations are fused, and from which they spring, is called eternity, *'olam*.'

While accepting the value of Jenni's basic sense of 'the remotest time', I should venture to suggest that at some points this is best supplemented with the sense 'perpetuity', or something like it. For example, for *'ebed 'olam* I think that 'a slave in perpetuity' is rather more convincing than 'a slave until the remotest time', as an account of the way in which the extant phrase came into existence. Conversely, the range of 'the remotest time' has to be qualified by the observation that certain contexts, which would seem natural from this rendering, such as 'in the remotest time, before man was created, such and such happened', do not occur for *'olam* at all, and as far as we can tell are impossible for this word. We might therefore best state the 'basic meaning' as a kind of range between 'remotest time' and 'perpetuity', and existing only in certain contexts at that.

Some further points must be added about the Hebrew usage here. Most important is to notice that in the sense 'the remotest time' no specification is given of how remote the time referred to is; precisions of this kind may be inferred from the context, but the word itself just tells us that the remotest time relevant to the present subject is meant. Thus in Josh. 24.2 the fathers had dwelt in a certain district 'from ancient times' or simply 'always' (*me-'olam*); Jer. 28.8 speaks of the prophets 'which have been before me and before you from earliest times' (*min-ha-'olam*). In these cases we might indeed translate 'from all eternity', but it would not raise any questions about an eternity which is other than time; it would mean, and would clearly be understood to mean, from remote times in the past. When a slave is *'ebed 'olam* (Deut. 15.17), he is a perpetual slave, not one held only on a contract lasting a few years. The phrase does not raise questions either of timelessness or of an eternity which is the unlimited totality of time. With certain other institutions *'olam* will mean permanence throughout the whole of time after a time of inauguration, e.g. *ḥoq 'olam* 'an eternal statute' or *bᵉrit 'olam* 'a perpetual covenant' (Ex. 29.28; Gen. 9.16). The question of the extent of time involved by relating an object to *'olam* is therefore relative to what the object is. In some cases, where the being of God is so related, nothing less than the totality of time would be meant; but this does not settle a question whether it would also involve some other kind of time or some kind of supra-temporality, and such questions can probably not be answered by appeal to the word *'olam* itself at all.

We may note before going farther that this sense of 'remotest time', whether we use the definite article or not, is itself more or less determinate from its own meaning, and thus furnishes another argument against Cullmann's argument from the plural *'olamim* or αἰῶνες. There is no such thing in classical Hebrew as 'an' *'olam* of which there may be several examples to string together. There may of course be *'olam* in two directions, remotest time or perpetuity in the future and in the past. These observations are equally true whether the definite article is used in Hebrew or not.[1]

When we turn to the Greek background, phrases with αἰών

---

[1] Cf. Jenni in *ZATW* lxiv (1952) 245 f.

in a sense of 'perpetually', 'eternally' are well established as far back as classical times.[1] In Hesiod we have ἀπ' αἰῶνος 'of old' as in biblical usage; cf. for example Luke 1.70. Ample evidence of usage can be given from later Attic writers like Lycurgus. Cf. also a good example at Isocr. 10.62. Of this material in general only one or two points need to be made.

(a) We may spare ourselves the common conflict which arises from trying to decide whether the Greek or the Hebrew background is decisive for LXX and NT Greek. This dilemma is often a false one. We may suggest in this case, as probably in many others, that it is the existence of a suitable idiom in the Koine which enables a Semitism to be expressed in Greek without undue strain.

On the other hand we see in this case the danger of an exclusivism which regards the circle of biblical Greek as alone important. It is most probably this assumption, which agrees with certain assumptions and procedures of modern biblical theology, which makes it seem natural for Cullmann, when discussing the bound phrases like ἀπ' αἰῶνος, to connect them with the free sense of αἰών actually existing in the New Testament, i.e. the sense of a particular age marked or defined by eschatological conceptions. The formation of the phrase ἀπ' αἰῶνος is not an innovation of New Testament Greek, and senses of αἰών existing in the latter are not necessary relevant for the elucidation of the phrase. In fact the ἀπ' αἰῶνος of Hes. *Th.* 609 has just the same sense, 'of old', as is found in the three New Testament cases at Luke 1.70; Acts 3.21; 15.18.

(b) It is true that the earlier history of the Greek word αἰών, in this unlike that of the Hebrew word, shows senses such as 'life' or 'lifetime' and 'long space of time',[2] and so in theory it might be possible for Cullmann to argue from this that some kind of exactly limited period is meant by αἰών in the phrases like ἀπ' αἰῶνος. But for one thing we have already seen that the sense here involved is the indefinite one of 'long space of time'. For another, it is not likely that Cullmann can appeal to the classical Greek development at all, for he is so active in re-

---

[1] See LS for ample evidence.

[2] I leave aside as irrelevant to our purpose such matters in the still earlier history of the word as are considered by R. B. Onians, *The Origins of European Thought* (1954), p. 200 f.

71

pudiating what he holds to be Hellenic conceptions of time that
he is not likely to invoke classical senses as a guide to the relation
of time and eternity in the New Testament. For yet another, it
is only within the Jewish-Christian eschatological schemes of
succeeding periods of time that there is any intelligibility for his
explanation of (let us say) εἰς τὸν αἰῶνα as 'into the age', i.e. as
into that time period which is limited on this side by its beginning now in the events of the Gospel story, and goes on on the
other side without limit into infinity. No one, surely, will regard
this as an intelligible explanation of the εἰς ἅπαντα τὸν αἰῶνα of
Lycurgus 106, or of other classical or Hellenistic passages.
Such phrases, like many parallel cases in the Bible, would be
more simply translated by phrases like 'in perpetuity', 'perpetually' or 'for ever'. Thus at Lycurgus 106 the reference
would probably be to the future of Sparta in perpetuity. From
some such cases we may see how it was possible for αἰών to be
adopted by Plato[1] for use in a sense of 'eternity' which was explicitly conceived to be changeless and timeless.

(c) One of Cullmann's main arguments was that αἰών was used
in the New Testament both for unlimited time ('eternity') and
for limited periods of time, and that therefore we can be sure
that in New Testament thinking eternity is the same stuff as
time except in that it is endless. But the facts to which he here
appeals in New Testament Greek are equally, or much more,
present in classical Greek, and in Plato in particular. In the
*Timaeus* (37 d) he explicitly says that his αἰών is timeless; but he
can elsewhere say that ἐμπειρία μὲν γὰρ ποιεῖ τὸν αἰῶνα ἡμῶν
πορεύεσθαι κατὰ τέχνην, ἀπειρία δὲ κατὰ τύχην 'expertness makes
our life proceed according to art, inexpertness according to
chance' (*Gorg.* 448 c), and can speak of those who are χαλεπὸν
αἰῶνα διάγοντας 'passing a hard life' (*Laws* 701 c). Thus Plato
uses αἰών not only for a timeless eternity but also for a limited
temporal period. If the existence of this duality of usage in the
New Testament proves that eternity is unlimited time, and not
something other than time, then the same can be proved for
Plato. But in fact Plato himself explicitly says that temporal
movement is absent from that which he called αἰών in the
*Timaeus*. We can certainly not, therefore, use the existence of a

[1] *Tim.* 37 d.

similar set of facts in the New Testament as proof that in it eternity is not other than time except by being unlimited.

Neither with the classical nor with the Semitic background in mind, then, can we admit that Cullmann has given a convincing case when he argues that the use of αἰών both for a limited duration and for the total unlimited extent of time is evidence that 'eternity' is the entirety of time, and nothing else. The way in which this argument distorts the facts may be seen from the summary of the usage of αἰών given by Cullmann:

'1. Time in its entire unending extension, which is unlimited both in the backward and the forward direction, and is thus "eternity".

2. The limited time which lies between creation and the eschatological drama, and which is thus identical with the "present" aeon or "this" aeon.

3. Periods of time which are limited in one direction but unlimited in the other, and in particular:

a. ἐκ τοῦ αἰῶνος, that period which lies before the creation, and which thus on the side towards creation has an end and thus a limit, but which in the backward direction is unlimited and unending, and which only in this sense is eternal;

b. The period which extends beyond the end of the present aeon, the αἰών μέλλων, which thus has its beginning, and thus a limit, in the so-called eschatological event, but on the forward side is unlimited and unending, and which only in this sense is eternal.'

Cullmann goes on to claim that this scheme shows how this 'naïve rectilinear conception of endless time' is the only one into which the redemptive history of the New Testament can fit.[1]

On the basis of our arguments already set out this scheme of Cullmann's can be criticized on the following grounds:

(a) The elements in the list are heterogeneous, and cannot legitimately be aligned in this way. Thus in particular points 2 and 3 b depend on the late Hebrew sense of *'olam*, and the corresponding New Testament sense of αἰών, as used in 'this age' or 'the coming age', i.e. of a particular defined period, while

[1] Cullmann, p. 41 f.; E.T., p. 48.

1 and 3 a have nothing to do with this sense. Cullmann's naïve scheme is in fact a homogeneous representation of his own view of time, with the four elements, namely the entirety of time, the period before the creation, the period from creation to the final events, and the period from the final events onwards to infinity. But only a most forcible effort can arrange the linguistic evidence within this scheme, and the 'naïve scheme' in fact demands extremely unnatural and artificial constructions of linguistic usage, and leaves other aspects of usage out altogether.

(b) A very obvious weakness is the treatment of the phrase ἐκ τοῦ αἰῶνος, with which we may suppose is also included ἀπ' αἰῶνος (cf. the passage cited above, p. 68). At Luke 1.70 we hear how God has spoken by the mouth of τῶν ἁγίων ἀπ' αἰῶνος προφητῶν αὐτοῦ 'his holy prophets from of old' (so RSV). Again at John 9.32 we have ἐκ τοῦ αἰῶνος οὐκ ἠκούσθη 'it has never been heard' or 'since earliest times it has not been heard'. Neither of these cases depends on the late Hebrew and New Testament sense of 'particular limited period'; but only an insistence on that sense can place them within Cullmann's point 3 a. Yet John 9.32 is the only case in the New Testament of ἐκ τοῦ αἰῶνος.

Furthermore, it is extremely obvious that the events mentioned in the New Testament with the temporal location ἐκ τοῦ (ἀπ') αἰῶνος do not fall into the period before the creation of the world, but into the period from the creation of the world up to the present for the writer. Thus in Luke 1.70; Acts 3.21 the prophets are ἀπ' αἰῶνος, but are certainly not intended to have been prophesying before the world was created, while the speaker in John 9.32 who said that something had not been heard ἐκ τοῦ αἰῶνος was probably not thinking at all of a time before creation, when the subject under consideration (the giving of sight to a blind man) can hardly have been very relevant. Thus even if a 'limited period' were here intended, it would not be a period before the creation of the world, but only a period limited by the present for the speaker or writer. This is of general importance, because it seems to be only from this use of ἐκ τοῦ αἰῶνος or of ἀπ' αἰῶνος that Cullmann concludes we have evidence for an αἰών or period before creation.[1] His interest in this αἰών before creation is surely the reason why he brushes aside the possibility

[1] See for this also p. 58; E.T., p. 67.

that the New Testament writers supposed time to have begun with creation.[1] But his evidence for this period, depending on his analysis of ἐκ τοῦ αἰῶνος, is really worthless.

In general there is a considerable likelihood that the early Christians understood the Genesis creation story to imply that the beginning of time was simultaneous with the beginning of the creation of the world, especially since the chronological scheme takes its departure from this date. Cullmann neglects, and indeed has to ignore, this likelihood, which could destroy his whole picture of the absence of qualitative difference between time and eternity. When he says that 'temporality is not in itself something human and given only along with the fallen creation',[2] we may well agree, but only ask if it is not something given with the creation before the fall, and indeed before the creation of man, which might be a natural way of reading Genesis 1. Cullmann says we cannot deduce from the ἐποίησε τοὺς αἰῶνας of Heb. 1.2 that time was something created, since αἰῶνες means 'world' here. Quite so; but it remains likely that Genesis 1 and other sources would be read as suggesting that time began with creation. This is the explicit interpretation of the creation story by Philo[3]: χρόνος γὰρ οὐκ ἦν πρὸ κόσμοι  ἀλλ᾽ ἢ σὺν αὐτῷ γέγονεν ἢ μετ᾽ αὐτόν. Perhaps Philo was in part inspired by Greek philosophy in his interpretation here, but even so it is an aspect of Greek thought which seemed to him to offer a reasonable account of the text before him, and which probably seemed sound to his Jewish contemporaries. One cannot therefore be sure that 'time is not bound to the creation'. The opinion, maintained by a dogmatician like Barth[4] and an Old Testament scholar like Eichrodt,[5] that time began with the first acts of creation, is surely a very natural way of interpreting the Genesis text and of stating how it would be read in the early Church.

This may be of some importance for the question of whether time might be conceived as coming to an end. Granted that it is likely that time was thought of as having a beginning, it is difficult to say absolutely that the late passage Slav. En. 65 is on

---

[1] Cullmann, p. 55; E.T., p. 63 f.
[2] Ibid., p. 55; E.T., p. 63.
[3] Philo, *Op. Mund.* 26. See in general S. Lauer, 'Philo's Concept of Time', *JJS* ix (1958) 39-46.
[4] *KD* iii/2. 525; E.T., p. 438.  [5] *ThZ* xii (1956) 122 f.

entirely non-biblical and non-Jewish ground in expecting that time would end.[1] Rev. 10.6 contributes nothing to a decision either way.[2]

(c) Similarly εἰς τὸν αἰῶνα in a case like John 13.8 or I Cor. 8.13 certainly refers to the future and thus by the scheme should certainly go into 3 b, but equally certainly is not a reference to, or dependent upon, the sense found in αἰὼν μέλλων. Again, the future time is counted, surely, not from the inauguration of the new age, but from the present at the time when the words are spoken.

(d) Another difficulty is that Cullmann's sense 1, that of the entirety of time, which to him is identical with eternity, is hardly found for αἰών in the New Testament except in certain combinations. Cullmann himself has to refer us primarily to the adjective αἰώνιος.[3] A sense of eternity or 'entirety of time' might also be found in certain genitive combinations, like τῷ βασιλεῖ τῶν αἰώνων (I Tim. 1.17), if the meaning here is not 'world', and very probably in Eph. 3.11, where κατὰ πρόθεσιν τῶν αἰώνων is commonly taken as 'according to the eternal purpose' (so AV, RV, RSV, Bauer). Some cases of εἰς τὸν αἰῶνα might also be taken as referring to the totality of time, thus for example John 8.35: ὁ υἱὸς μένει εἰς τὸν αἰῶνα, in which 'the son remains eternally' might be taken as at least one of the suggestions made by the writer. The same might be said of I Peter 1.25 (approximately = Isa. 40.8): τὸ δὲ ῥῆμα Κυρίου μένει εἰς τὸν αἰῶνα. Here again it may be reasonable to suppose that the reference is one to the entirety of time. With these we can compare such examples as:

---

[1] Mentioned by Sasse in *TWNT* i. 202; Cullmann, p. 39 n.; E.T., p. 46 n.

[2] Modern interpretation has shown a strong tendency to take this as 'there should be no more delay' (so RSV). It is true that χρόνος sometimes amounts to 'delay' when it is the object of certain verbs, but this hardly proves this sense for χρόνος οὐκέτι ἔσται, and actually the arguments for 'delay' here have been built largely upon the verb χρονίζειν, which certainly means 'to delay'; but this is hardly evidence for χρόνος itself. All the ancient versions translated with words for 'time' here, although they had words for 'delay', and one feels they were hardly to be blamed. It is rather the context within the Apocalypse as a whole that shows that the sense here is not an abolition of time and its replacement by timelessness, but 'no more time' from the words of the angel until the completion of the divine purpose. This certainly comes near to 'delay' in general effect; it remains uncertain that 'delay' was what the angel said. For the arguments for 'delay' see in particular E. J. Goodspeed, *Problems of NT Translation* (Chicago, 1945), p. 200 f. The analogy of Hab. 2.3 and Heb. 10.37, quoted by Cullmann, p. 42; E.T., p. 49, is a use of the verb χρονίζειν, and is no analogy.

[3] Cullmann, p. 41 n.; E.T., p. 48 n.

Matt. 21.19 οὐ μηκέτι ἐκ σοῦ καρπὸς γένηται εἰς τὸν αἰῶνα
I Cor. 8.13 οὐ μὴ φάγω κρέα εἰς τὸν αἰῶνα

In these the sense is certainly not of the whole of time, but of the future from the time of speaking. We can perhaps say then that εἰς τὸν αἰῶνα is capable of bearing either meaning, that of 'eternally' or 'throughout the whole of time', or that of 'always from now on', or negatively 'never'. But the fact that the phrase can be used with either of these meanings surely does nothing to show, as Cullmann thinks, that 'eternity' is only unlimited time. It shows rather that these two possibilities are left open when the phrase is used, so that when used with certain subjects it will be understood to refer to the totality of time (if not a supratemporal being), while with others it will be understood to refer to a future limited or unlimited. The same Greek phrase, then, may be used firstly for the totality of time, and secondly for a perpetuity in some state for the whole of a limited period, and negatively for the continual avoidance of a particular action for the whole of a limited relevant period (as I Cor. 8.13; Matt. 21.19). It does not follow that there is no essential difference of quality between the states referred to in the first reference and the second.

(e) Cullmann rightly sees that for his sense 1, the entirety of time without limitation, he has to turn mainly to the adjective αἰώνιος. But here there are other difficulties for his theory. Firstly, this adjective is not used with the sense of 'lasting for a limited period or aeon', so that the argument which he used with the noun, namely that the same word is used both for the entirety of time and for a limited period, does not work here. The sense of αἰών found in 'this age' has no adjectival counterpart. Secondly, the cases of αἰώνιος refer fairly uniformly to the being of God or to plans and realities which, once established by him, are perpetual or unchanging. Since the word is not used of more mundane realities like the flowering of fig-trees, one cannot argue that the same kind of temporality is attributed to these as to the being of God. Thirdly, I do not see how Cullmann has not contradicted his whole case when he says of this adjective that 'since eternity in this sense comes into consideration only as an attribute of God, the adjective αἰώνιος has the tendency to lose its temporal sense, and is used in the qualitative sense of divine-imperishable'.[1] The admission that there is a sense of 'eternity' which can also be a 'qualitative' kind of imperishability, and therefore something other than and more than mere unlimitedness of time, is surely the main point against which Cullmann has generally been arguing.

Thus when Cullmann, in summarizing the argument again, says that in the New Testament 'above the distinction of "time" and "eternity" there stands the *one* temporal concept of the aion, which embraces both',[2] we have not only a case of the faulty use of 'concept' already criticized, but an ignoring firstly of the variety of different syntactical contexts in which αἰών and αἰώνιος are used, and secondly of the fact that the phrases are used of the being and action of numerous different objects, for only some of which (and in particular God and his deeds) the sense of perpetuity throughout the whole of time would be considered; while for these cases (such as the being of God) the sense of continuance through only a limited period would not be considered. The argument ignores also the basic fact that the cases involving αἰών in senses like 'the present age' can be easily distinguished from, and are in no way dependent upon, its sense as 'perpetuity'.

One other point made in Cullmann's treatment of the Greek terminology for time is worth discussing, for it is a statement typical of the mood and the methods of much modern interpretation. He assures us[3] that 'none of the temporal expressions of the New Testament, not even χρόνος, has as its object time as an abstraction'. The contrast between a Hebraic concreteness of thought and a Greek abstractness is a commonplace of present-day theology. What is harmful, even if this contrast is a valid one, is the thoughtless transfer of it into statements about linguistic usage. In this case the transfer simply produces a meaningless statement. Is there a word in classical Greek which *means* 'time as an abstraction'? Is there in any language? If not, what is the point in emphasizing that Jewish and Christian usage has no such word? The way in which the Greek philosophers

---

[1] Cullmann, p. 41, n. 21; E.T., p. 48, n. 21.
[2] Ibid., p. 54; E.T., p. 62.
[3] Ibid., p. 42; E.T., p. 49.

(or some of them) thought about time may or may not be reasonably described as an abstract one, but this does not mean that they had a word *meaning* 'time as an abstraction', any more than Cullmann has in his German or we in our English.

Thus, to put the point in another way, the sense of 'time as distinguished by its content', which is so often picked out as the characteristic one for Hebrew *'et* and for other words in biblical usage, is (in so far as it is a meaningful description of the semantics of a word at all) perfectly easy to substantiate from classical Greek usage of χρόνος. The average Greek speaker using χρόνος was not talking about 'abstract time' or 'time as an abstraction' but about some time in which something was happening, or some time the elapse of which was important for the understanding of the description of some event. 'Time as an abstraction' is no fair statement of what was in the mind of the envoy in Aristoph. *Ach.* 137-40, who stated that

χρόνον μὲν οὐκ ἂν ἦμεν ἐν Θράκῃ πολύν
'we wouldn't have been a long time in Thrace, but . . .'
  and
τοῦτον μετὰ Σιτάλκους ἔπινον τὸν χρόνον
'I was drinking with Sitalces all this time'.

Nor was it in the mind of Thucydides when he wrote (ii. 26) that ὑπὸ δὲ τὸν αὐτὸν χρόνον τοῦτον ᾿Αθηναῖοι τριάκοντα ναῦς ἐξέπεμψαν περὶ τὴν Λοκρίδα. 'About this same time the Athenians sent out thirty ships round Locris.'

When philosophers like Aristotle came to talk about the reality called time and to produce that understanding of it which modern theologians might call 'abstract' in comparison to Hebraic thought, they used this same word χρόνος. They did not need to have a word meaning only 'abstract time' or 'time as an abstraction', and this was just as well, for no such word was available. It is clear therefore that Cullmann's claim that χρόνος in the New Testament did not mean 'time as an abstraction', a claim clearly made in agreement with his continual effort to set biblical thought in contrast with Hellenic, is a meaningless statement.

We may here conclude our study of Cullmann's argument from the New Testament vocabulary stock. Readers should perhaps be reminded that I am by no means trying to prove that Cullmann is wrong in his contention that there is no qualitative distinction between what we call 'time' and what we call 'eternity'; I am showing only that this contention cannot be proved by Cullmann's study of the vocabulary, and that his study of it contains many faults and inaccuracies. In so far as Cullmann's arguments elsewhere in his book depend on the terminological examination, they are likely to be faulty. The basic fault in the whole procedure is the assumption that the vocabulary stock is laid out in a pattern which correlates exactly with the mental pattern of New Testament thinking about time. This fault is one that Cullmann shares with his critic Marsh, and they both use similar general procedures in handling the Greek vocabulary, although their general opinions about time are widely different; thus Marsh's criticisms of Cullmann, although often noticing real points of weakness, are unable to discern the basic errors in method, and often offer equally erroneous suggestions to replace those of Cullmann. A survey of the books here criticized, and of the discussion provoked by them, leads strongly to the conviction that recent theology has become extremely uncritical of the semantic methods used in argument from linguistic evidence. The discussion has not, as far as the writer has found, included a critical sifting of the use of linguistic evidence, and has generally been a discussion simply of the general position about time and eternity propounded.[1]

We here conclude our close study of Cullmann's section on terminology. Further comments on Cullmann's position in more general respects will be made later. It remains to remind readers that the section on terminology is basic for much of what follows in Cullmann's book, establishing as it does the two great assertions, that time is a line and that it differs from eternity only in that the latter is unlimited. On the other hand much that Cull-

---

[1] Even Bultmann's very critical review of Cullmann in *ThLZ* lxxiii (1948) 659 ff. touches on only a few linguistic points. For other discussion of Cullmann see W. A. Whitehouse in *Theology* l (1947) 230 f.; T. W. Manson in *JTS* xlviii (1947) 221 ff.; J. A. T. Robinson in *SJT* iii (1950) 86 ff.; T. F. Torrance in 'The Modern Eschatological Debate', *EvQ* xxv (1953) 175 ff.; K. G. Steck, *Die Idee der Heilsgeschichte* (Zollikon, 1959), p. 43 ff.; J. Frisque, *Oscar Cullmann: Une théologie de l'histoire du salut* (Tournai, 1960). Of Marsh: W. Manson in *SJT* vi (1953) 308 ff.; J. A. T. Robinson in *Theology* lvi (1953) 107 ff. Of Robinson: W. A. Whitehouse in *SJT* v (1952) 313 ff.; R. H. Fuller in *Theology* liv (1951) 269 f.

mann says later will not be criticized here, and our purpose is
no more than to point to the precariousness of such results as
depend simply on the procedures criticized here. Our criticisms
are emphatically not of Cullmann's scholarly method as a whole,
but only of that method as applied to the subject of time through
a terminological and lexical study. His work on Christology,
where the material is organized according to the various Messianic
titles and designations, is on a different footing altogether.
These different designations have roots in different religious
traditions, and the validity of an attempt to work out a Christology by relating them to each other is quite different from the
validity of an attempt to work out the nature of time and eternity
from a lexical structure of the words for these realities.[1]

[1] I regret that this book was already set up in type before I discovered that some
of my arguments had been anticipated, and some others provided, by A. L. Burns
in *ABR* iii (1953) 7-22. Although I do not agree with all of the points made by
him, his basic criticism of the use of the καιρός-χρόνος distinction is acute and
correct. Other treatments of our subject will be found in the works of J. Schmidt,
*Der Ewigkeitsbegriff im alten Testament* (*Alttestamentliche Abhandlungen* xiii.5, Münster,
1940), R. Martin-Achard in *RThPh* iv (1954) 137-40, E. Perry in *JBR* xxvii (1959)
127-32, and P. L. Hammer in *JBR* xxix (1961) 113-8. I have not seen the Kiel
dissertation of W. Vollborn, mentioned by Martin-Achard, p. 138. Cf. also J.
Mánek in *New Testament Studies* vi (1959-60) 45-51.

I should also add another clear case of καιρός used for a period by the Hellenistic
Jewish poet Ezekielos, who writes at line 32:

ἐπεὶ δὲ καιρὸς νηπίων παρῆλθέ μοι
ἤγαγέ με μήτηρ βασιλίδος πρὸς δώματα ...

The words καιρὸς νηπίων mean 'period of infancy'. In his edition in 1931 J. Wieneke
commented correctly: καιρός scripsit poeta pro χρόνος, quam consuetudinem
posterioris scriptores temporis sequuntur. He went on to add illustrations from
Polybius and the papyri.

# IV

## HEBREW WORDS FOR TIME
## AND ETERNITY

W E now turn to discuss some questions in the Hebrew vocabulary relating to time. Here we must begin with some reference to a book now almost 100 years old, but one which probably still has some considerable influence on what is said about time by theologians, even where its direct influence is unacknowledged or imperceptible.[1] Many statements still being made in theological discussions about time can be traced back as far as Orelli, and it probably does not much matter whether they can be traced farther back or not. The work of Orelli represents the taking up of the subject on the new basis of mid-nineteenth-century philology, and thereby the introduction of standards which were new in comparison with the older works of Hebrew synonymists and with the traditional interpretations of Christian orthodoxy. Orelli's first great contribution was that of a constant historical and comparative perspective to the subject (as the sub-title of his book proclaimed), and neither theologians nor other scholars have been disposed to hold that this approach can be by-passed and earlier methods resuscitated. His second contribution was the integration of this historical-comparative material into a framework of general observations of a psychological and philosophical kind about the nature and growth of language; these observations have a strong kinship with what has continued to be an understanding of language widely held among theologians, so that it will be worth while to discuss them at some length.

It is no disparagement of Orelli's book to maintain that it is now entirely antiquated. It has not however been replaced by a more modern work, except over part of the area, namely the usage of the word *'olam*, the most important by far of the words treated under 'eternity'; here it has been wholly superseded by the careful and scholarly work of Jenni; this does not however go in detail beyond the end of the Old Testament period. Whether

---

[1] C. von Orelli, *Die hebräischen Synonyma der Zeit und Ewigkeit genetisch und sprachvergleichend dargestellt* (Leipzig, 1871). Cf. Jenni in *ZATW* lxiv (1952) 197 f.

or not it is desirable that the whole group of Hebrew words for 'time' should be handled together afresh, so as to cover Orelli's ground again in a more up-to-date way, I shall attempt nothing of the kind. The purpose I have here is only to bring out certain points where Orelli's judgements in detail may have excessively influenced later interpretation, and certain aspects of his method, the criticism of which can help to lead us to more accurate methods in handling semantic questions.

Firstly, something must be mentioned for which Orelli can of course not be blamed, the rudimentary character of the historical and comparative philology current at a time when the grammar of Accadian was still most imperfectly known, when the discovery of Ugaritic still lay far in the future, and when in particular the phonological correspondences between the various Semitic languages could be stated only with considerable imprecision. This last fact naturally makes specially precarious the method of comparative etymology which was so important for Orelli. Thus for the derivation of Hebr. *'et* 'time' Orelli first considers *'ada* 'pass by' to be phonologically possible[1]; then he says that a better case can be made for a derivation from *'ana* 'meet'[2]; but finally he prefers one from *ya'ad* 'appoint',[3] which however in his opinion is in respect of sound close to yet another verb, *'adad* 'count', from which we have the Aram. *'iddan* 'time', commonly used to translate *'et.*[4] Where so many etymologies are taken to be phonologically possible, it means that much weight has to be given in choosing between them to a rather general semantic method which by reasoning from the actual nature of time works out likely ways in which the words might have been developed. This in fact is very apparent in Orelli's work. One more example of comparative semantics which would to-day be considered precarious may be given: Orelli adduces the root *tan* of τάννμαι, τείνω, and τιταίνω to help in the elucidation of Hebr. *'etan* 'perpetual'.[5]

Secondly, it is clear that Orelli started from a number of significant philosophical and general cultural presuppositions, which furnish him with much of his general method and of its applications. Thus his choice of time as a subject has its origin in a philosophic opinion. 'Time is not an object, but is like space

[1] Orelli, p. 17.      [2] Ibid., p. 18 f.      [3] Ibid., p. 47 ff.
[4] Ibid., p. 53.       [5] Ibid., p. 94.

a mere form of existence, yet lies even farther away from sense perception than the latter.'[1] Only philosophical investigation can disclose what time really is, yet some awareness of it must have been present in the earliest beginnings of human thought. These considerations make it seem specially valuable to undertake an investigation of the linguistic expression of the awareness of time. Two of the philosophical positions which Orelli accepted should be mentioned in particular. Firstly, the priority of sense perception to the knowledge of what is more intangible or 'mental', so that the latter is known by a process of 'apperception' which is mediated through the knowledge of that which is perceptible by the senses.[2] Secondly, the impossibility of coping with time in any philosophical system without linking it as closely as possible with movement;[3] thus the ὁ χρόνος ἀριθμός ἐστι κινήσεως κατὰ τὸ πρότερον καὶ ὕστερον of Aristotle appropriately forms the motto to the chapter on time.

Thirdly, however, and most important for our subject, these philosophical judgements are also used as, and interwoven with, judgements about the nature, growth and history of language, and of these the following should be mentioned in particular:

(a) It seems obvious to Orelli that the older and simpler the word for something is, the more ancient historically is the human knowledge of that thing. Objects and notions, the words for which are obviously derived from other words, must have become known later. Being 'comprehended with greater difficulty and learned of later,' they must also have been 'named more artificially and later', than those other objects which 'lay nearer to the perceiving and knowing mind'. And in particular, it is unmistakable that the objects learned of and therefore named earlier are the objects of sense perception. This general rule can be observed more or less in all languages 'as it has its basis in the universal natural laws of human development'[4]; it is a well-tried principle of classical philology that 'notiones verborum propriae omnes sunt corporeae sive ad res pertinentes quae sensus nostros externos feriunt'. But in Orelli's opinion the correctness of the rule is much more clearly demonstrated in the Semitic languages than elsewhere. This is because the organic sub-division of the individual word-families is very easy to discern in Semitic, and

[1] Orelli, p. 8.     [2] Ibid., p. 5.     [3] Ibid., p. 13 ff.     [4] Ibid., p. 4.

84

when it is traced as far back as possible it almost always leads to a meaning referable to sense perception.

(b) The process of 'apperception' is not only a process in which knowledge of intangible realities is gained through the mediation of the tangible, but also a linguistic process through which the sense of words is changed. For the intangible realities now coming to be known new names were not devised, but names of tangible realities which seemed to have some similarity to, or to give some reminiscence of, intangible realities, were used or adapted for the latter. According to Orelli homonyms and synonyms are produced by this process. Homonyms appear when different intangible realities are named by the same word which originally meant a tangible reality, as when Hebrew *'amad* 'stand' is used both to mean 'continue' and to mean 'stop'; similarly *'ada* 'pass by', which will be of importance for Orelli's study of words for time, is said to mean both 'pass, pass away' and 'go on, continue'. Synonyms on the other hand appear when one intangible idea is represented by adaptation of two or more tangible realities. An illustration of this is 'eternity', where in Hebrew we have the synonyms *'olam* and *neṣaḥ*. The former is from *'-l-m* 'hide' and thus indicates the time which is hidden from human perception or comprehension; the latter is from a root meaning 'shine, be bright', and indicates that which is brilliantly superior to all limits.[1] It follows that the more inventive a people is in the use of its imagination upon a matter, the richer its vocabulary will become in synonyms designating that matter. Orelli thus later maintains that Hebrew is remarkably well provided with words for eternity, and that this furnishes an interesting and significant contrast with both Arabic and Greek. Arabic has many more words than Hebrew for time, while the reverse is true for eternity. Greek likewise is poor in designations for eternity, and for the LXX could offer no more than the poor αἰών, 'which does not have this meaning either originally or exclusively'. These circumstances can be certainly attributed, in Orelli's opinion, to the effect of the pure idea of God, as known in Israel, upon the terminology for eternity.[2]

It remains true, however, for Orelli that the study of the synonyms which thus originated is specially valuable, because,

[1] Orelli, p. 97 f.     [2] Ibid., p. 105.

even when the reality named is the same, each of the synonyms reveals a different way by which the mind has come to the subject. It is regrettably to be admitted that the linguistic consciousness commonly ceases to observe this past history and indeed becomes quite unaware of it; but this is only all the greater reason why the philologist, 'part of whose task it is to develop psychologically the history of concepts from language', should investigate it. The plan of Orelli's book, with its grouping together of the synonyms for time and eternity, the tracing of each word back to a root meaning, and the attempt to explain from this root meaning how it fitted into the apperception of the reality designated, is thus simply a carrying into effect of these considerations.

(c) It follows that the linguistic investigation is dominated by *etymological* methods, for such alone can perform the linguistic task as prescribed within the general framework of Orelli's presuppositions. In the circumstances this can hardly fail to mean some loss of attention to actual usage. In the case of synonyms derived from different etymological origins, as discussed above, the linguistic consciousness, unaware of the nuances arising from the etymologies, betrays itself as dull and lacking in subtlety,[1] and one has the impression not seldom that a philology of Orelli's kind regarded actual usage as a sort of degeneration whenever it departed from or ceased to remember etymological origins.[2] Even at the highest estimate of Orelli's results, then, all they give us is a kind of prehistory of Hebrew vocabulary and not an account of its actual usage in the period of the texts. The giving of a synchronous account of usage within one period lay beyond Orelli's horizon. Thus for the three words for 'eternity' Orelli can form a scheme where '*olam* is 'the time, the limits of which are not perceptible or not existent', '*ad* is 'the time which carries on to the uttermost thinkable limits', and *neṣaḥ* is 'the time which transcends all limits'; but he has himself to admit that in usage little essential difference of sense seems to have been known.[3] In fact the linguistic consciousness of the writers was 'dull' towards the etymological nuances, and no wonder, since much of the pre-history of the words must have been unknown to them.

---

[1] Orelli, p. 7.  [2] Ibid., pp. 19, 48, 52, 57, 73 f., 80, 105.
[3] Ibid., p. 98.

Even where we admit as far as possible a justification for Orelli's emphasis on etymological method, a difficulty in applying it is evident. Even where, as often happens in his discussion, a word for 'time' can be reasonably probably assigned to a root which in an extant verb means 'meet' or 'appoint', there is still a rather large blank area in which there is no direct evidence for the way in which the form and the sense 'time' have developed from this root. There is a danger that this gap may be filled in by a suggestion, the main force of which lies in its conformity to the general theory of time and of linguistic origins which are held in the discussion as a whole. Thus Orelli interprets Hebr. *ša'a* 'hour' as derived from the root seen in Arab. *sa'ā* 'go about, run', an interpretation which fits in with his general theory that words for shorter periods of time were derived from words for faster spatial motions.[1] Such general theories are a danger for etymological method rather than a support for individual solutions.

(d) I have already mentioned that the connection of time and movement was philosophically axiomatic for Orelli. This is not only philosophically true but linguistically important: 'all languages express this connection'.[2] We speak of time 'passing' or 'coming', of a *decursus aetatis* or a *fluxus temporum*. Not only so, but we find once again an 'apperception' between space and time, in which the latter is 'apperceived' by means of the former. Here once again an epistemological priority of space is assumed without question to involve a linguistic history in which temporal words are derived from spatial. The illustration given is the use of words like 'before' and 'after' for temporal as well as for spatial relations. Here however Orelli has to go on to remark a considerable inconsequence not only in Indo-European languages but also in Semitic. Thus for the future one says in Arabic that which lies before, *'al-mustaqbil*; but in another Semitic usage, exemplified from Hebrew, what is 'before' is the past, *qedem* or *re'šit*, while the remotest future is the 'farthest behind', *'aḥªrit*.[3]

---

[1] Orelli, p. 24, cf. p. 26.     [2] Ibid., p. 13.

[3] This part of Orelli's argument is criticized by T. Boman, *Hebrew Thought compared with Greek*, p. 130, cf. p. 149 and my criticisms of Boman's argument in *Semantics*, p. 75 ff. While I think that Orelli's argument deserves some criticism here, Boman builds up an even worse systematization of the relation between linguistic phenomena and ways of thinking about time. Orelli admitted an inconsequence in the phenomena here, which to his credit he did not try to iron out. Boman does

This is an ambiguity which arises when opposed metaphysical concepts attach themselves to the same physical concepts. The difficulty at any rate does not make Orelli feel that he has not shown the dependence of temporal upon spatial terms linguistically.

At any rate Orelli goes on to study the development of temporal words from words meaning movement of some kind. For Indo-European many words are derived from the simple root *i* 'go'. The same thing happens in Semitic, but there is a large variety of words taken from different kinds of motion, fast and slow, sudden and abrupt, agitated, circular and so on, each of which goes to bring out some particular 'modification of the time concept'. This general conception forms the frame for the first part of Orelli's study of words for time. Later however he goes on to maintain that alongside this group of words, which represent time as a phenomenon affecting or encountering man, there is another group in which time is characterized as that which is fixed, determined or ordained.[1] The contrast of these two groups forms his summing up of words for time.[2] It is not quite clear how the second group fits into the more general philosophical framework which I have described.

(e) Orelli also uses philosophical presuppositions for the relation of the finite and the infinite, presuppositions which later naturally become of great importance in the discussion of the relation between time and eternity, which is a natural consequence of taking up the subject in the way he has done. Language tries to express with its undeveloped resources things which even a developed intelligence cannot master; and in particular the infinite is something of which an awareness exists in the world of ideas, long before it is clarified by thought. Language tries constantly to find an adequate expression for the infinite, and its lack of success in this drives it to ever new attempts, which have value because of the peculiar light they throw on the ways in which mankind tries to reach out to the infinite from the finite

iron it out, maintaining that the apparently discordant linguistic phenomena belong to two different mental attitudes, with which they are respectively in complete concord; and for Hebrew in particular he believes that no temporal terms are derived from spatial, and that what is future is never represented by words suggesting what lies 'before'. For more general criticisms of Boman's picture of Hebrew thought about time, see *Semantics*, pp. 72-80.

[1] Orelli, p. 45.      [2] Ibid., p. 62 ff.

and sensible.[1] From the beginning the study as planned is to be divided between 'finite and infinite time, or time and eternity', although we are warned that the difference is not an absolute one.[2]

The detailed discussion of words for eternity begins with the statement that 'language raises itself from the sensible to the spiritual, from the finite to the infinite'.[3] For further application of this generalization Orelli takes over from traditional theology, and applies cheerfully to linguistic change, the terms *via negationis* and *via eminentiae*. The former reaches the infinite by denial of finitude. Linguistic examples would be Gk. ἄπειρος and Lat. *infinitus*. This kind of linguistic *via negationis* will not be found in Hebrew, where compounds formed with negative particles can hardly be said to exist. Nevertheless there is a certain kind of *via negationis* involved in the chief Hebrew word, *'olam*. Derived in Orelli's opinion from '-*l-m* 'hide',[4] it means 'that which is hidden', the time which was removed beyond human sight and which therefore could no longer be perceived by man. There is however an important difference in the *via negationis* here, for the formation does not imply an absolute denial of limit, as is the case with ἄπειρος and *infinitus*, but a denial of limit 'to a certain extent only in an epistemological sense'. The limit cannot be known by man, but the way in which the word is derived indicates that man can pass from perception or experience to the conception of the unlimited, and that the unlimited in this case is not totally other than the limited.

The *via eminentiae* on the other hand reaches the infinite not by the denial of the finite but by its enlargement, prolongation or intensification. This process is represented in Hebrew by *'ad*, derived by Orelli from *'ada* 'pass by', which represents infinite time by continual extension beyond all limits. In another form it is represented by *neṣaḥ*, which starts from a sense of 'brilliant light' and from thence comes to mean 'excellence, superiority, victory'; the word thus designates that infinite time which exceeds and excels all ordinary time.[5]

Nowhere then does the philosophical-philological parallelism which pervades Orelli's method become more clear than in his

---

[1] Orelli, p. 8.       [2] Ibid., p. 9.       [3] Ibid., p. 67.
[4] Ibid., p. 69 ff.       [5] Ibid., pp. 86 f., 95 ff.

use of the terms *via negationis* and *via eminentiae* for the process of derivation of words designating infinite realities.

This then may suffice as an account of the philosophical pre-suppositions which seemed obvious or reasonable to Orelli, and of the way in which the philosophical positions were simply transferred into, or interwoven with, supposedly historical processes of linguistic origin and change.

Apart from the general interest in examining the principles of Orelli's work, its importance for us lies in the numerous detailed statements about the biblical words for time which continue to appear in modern studies, and which either were already found in Orelli or are closely allied to statements of his, and secondly in the extent to which general aspects of his method have been perpetuated in more recent theological work.

Among the detailed statements still appearing in discussion which appear to go back to Orelli are the following: (a) Hebrew has no word for 'time as a quite general notion' (p. 64); (b) Hebrew has no word corresponding to Greek χρόνος, and *'et* corresponds to καιρός (pp. 18-9, 48 ff., 58 f.); (c) the use of compound expressions like *lᵉ-'olᵉme-'ad* is an attempt by language to express the incomprehensible (p. 81); (d) the use of the plural *'olamim* indicates a sense of 'age, period' for *'olam*, several of which ages or periods are expressed by the plural (p. 81 f.)—exactly the argument used for the plural αἰῶνες by Sasse and Cullmann; (e) the Greek αἰώνιος means 'belonging to the αἰὼν μέλλων' (p. 85); (f) it is significant for the idea of time that the Hebrew verb system does not express the temporal differences of past, present and future (p. 63 n.).

To these we should add one or two more general results as stated by Orelli, again in forms which have remained influential: (g) The Hebrew words for time, whether they have arisen directly from words for movement, or whether they have arisen from determining processes in which time is named as determined by human needs, have this in common, that 'they characterize time not as a pure continuum, as a universal entity or indeed as an abstract form, but as something as individual and concrete as possible' (p. 63). (h) Since words for infinite time or eternity are derived by use of words already referring to temporal movement, whether by extension, by cessation, by persistence, or by reaction, in relation to such movement, we can say that 'eternity is thus here understood throughout not as something which stands outside or above time, but as a continuum into which time runs out' (p. 99 f.).

Returning to the detailed points, namely (a) to (f), some of them, or at least some very similar arguments, have already been criticized above. For (a) and (b) some remarks remain to be made, and we shall turn first of all to (b). We have seen that the idea that Hebrew has no word meaning χρόνος is one on which a great deal was based by Marsh in his arguments about time. Against this one must say the following:

Firstly, Orelli's statements about καιρός and χρόνος, and about the possibility of finding corresponding words in classical Hebrew, are made wholly without reference to Hellenistic or New Testament usage, of which in this connection he made no investigation, as indeed he had no call to do (although it is true that occasionally he refers to New Testament usage in other connections). For the distinction between the two Greek words Orelli quotes several passages which distinguish the words, but all are from glossators and commentators of extremely late date (5th-6th century A.D.), who merely reproduce the classical usage. The large area in which the two Greek words were non-contrasting in Hellenistic Greek was not noticed by him. His statement that '*et* corresponded to καιρός and not to χρόνος has no relevance for New Testament usage and is positively misleading when used (as by Marsh) with reference to it.

Secondly, Orelli here, as not infrequently, made other statements which modify very considerably the categorical assertion which we have just discussed; this is a not infrequent event in Orelli's arguments, because he often makes a statement on the basis of his favourite ground of etymological investigation and then goes on to admit that usage did not coincide with what he has been saying. In this case, though he believed that the sense of '*et* was 'a peculiarly defined time' with reference to events occurring in it and thus corresponding, as he said, to Greek καιρός, he explicitly states that, since Hebrew 'had no other word at all for the simple concept of time', this word, '*et*, can be used in cases of abstract duration (*die abstrakte Zeitdauer*) where χρόνος would be the corresponding Greek (p. 19). On p. 52, after another treatment of '*et*, in which he understands it as 'determined time' or 'definite time' (for something), he says that the word could be very easily widened to a sense of 'time' in general 'because all time is more or less determined by its content'. And he goes on to argue that, 'since Semitic antiquity did not need an entirely abstract word, because it did not have the abstract concept of time', '*et* in classical Hebrew could suffice for the concept of 'time'. Unfortunately he does not cite examples for either of these arguments. Once again, finally, he mentions that *zᵉman*, while in his opinion 'properly' meaning something 'definite and quantitative', later in usage lost much of its original definiteness, and is used in later Hebrew for 'the philosophic concept of time as an abstract form', just as in Syriac it is almost the only word for time and in Arabic has lost the express character of definiteness. We can conclude then that Orelli by no means supposed his assertion that '*et* corresponded to καιρός and not to χρόνος (in classical senses) to be valid for the whole of classical Hebrew usage; but unfortunately we can do no more than surmise which passages he had in mind as exceptions.

Marsh's statement that Hebrew 'cannot translate χρόνος' is therefore hardly uniformly supported by Orelli himself. For further empirical evidence whether this Greek word could be translated into Hebrew we must look in vain, unless we are willing to go down to mediaeval usage. It is however surely not wholly irrelevant that χρόνος is used not infrequently in the LXX in translation of Hebrew words, a point made by Millar Burrows in his criticism of Marsh.[1] Putting the question in another way

[1] In 'Thy Kingdom Come', *JBL* lxxiv (1955) 1-8; see especially p. 4. This is so, even though it in itself does not *prove* Marsh to be wrong, as Eichrodt says (*ThZ* xii (1956) 107 n.). Burrows was entirely right to make the point. If the LXX had in fact never used χρόνος, we should have been told repeatedly that this showed how 'Greek temporal concepts' or 'the chronological idea of time' could not be used in translating the Old Testament; and it is only right to point out that χρόνος was in fact used quite often. Eichrodt's objection also depends partly on the usual confusion of word and concept. Whether there was a different Greek 'time concept' or not, there is every reason to say that numerous instances of the Greek word χρόνος in classical passages would be naturally translated into classical Hebrew with '*et*.
On the other hand, it must be admitted that Burrows' criticisms of Marsh are at times very indiscriminate, and contain much with which the present writer could not agree. Moreover, Burrows does not notice the really fatal weakness in Marsh's case, i.e. the fact that καιρός is often used of time in general or of periods of time.

which can hardly be answered empirically but is surely not with-out validity, we can surely translate into Hebrew many passages in which χρόνος appears, say in a classical historian like Thucy-dides, in some cases using *'et* and in others probably *yamim*.[1] The idea that χρόνος cannot be translated into Hebrew rests probably not on examination of actual passages where this word is used, but on the idea that it is a word meaning 'time as an abstraction', as already discussed above. The point (a) in our survey of Orelli, namely the assertion that Hebrew had no word for 'time as a quite general notion', is already considerably affected by these observations on the point (b) about the translation of χρόνος, but further observations about it must wait until a little later; for the moment we turn to the more general results which have remained from Orelli's work, as stated in (g) and (h) above.

The criticism of the general results (g) and (h) is the simple one that these results were almost bound to emerge, because the philosophical axioms which Orelli assumed in the beginning, along with the assumption that philosophical assertions about the process of knowledge were also philological assertions about the process of deriving words, made it very difficult for any other results to be possible. Thus it is a result that the words for time do not characterize it as a pure continuum or as an abstract form; but what else could be expected, when it was assumed in the beginning that time must be related to movement and that accordingly words for time would naturally be derived from roots designating some kind of movement? Again, it is a result that eternity is not something above or outside time; but what else is possible when it has been assumed that language moves from the finite to the infinite and bases its terms for the latter upon the former, and that the process of linguistic derivation reflects more or less exactly the process of knowing the subject? Only a very obvious and awkward piece of evidence, such as a word for 'time' that had nothing to do with any kind of movement, or a word for 'eternity' that had no connection with any finite and tangible reality, would have seemed sufficient to upset Orelli's scheme.

The general nature of Orelli's work, then, was such that we have to use his results with great caution, in spite of the careful observation and general erudition which has gone into them.

---

[1] See BDB, s.v. *yom*, sections 5 and 6, p. 399.

His manner of interweaving philosophical generalizations and linguistic processes, and regarding statements about the one as automatically transferable to the other, is not merely a habit for which allowance must be made; it is the reason for the writing of his book in the first place, and the force which has dictated the plan of it. The book was never intended to be merely a survey of usage in the Old Testament of words in the area of time and eternity; it is a designed structure in which the words, as interpreted by the overwhelmingly etymological method, a method used precisely because it was understood to present the actual sequence of the growth of knowledge, are set in position in a framework set by the express philosophical presuppositions. This is the reason why it is unlikely that Orelli's work can ever be brought up to date. We may write afresh about words for time and eternity, but can hardly share Orelli's faith that when they are brought together they will fall into a pattern symmorphous with the truth about time and eternity, and with the *ordo cognoscendi* in which these realities have penetrated the consciousness of man.

Orelli's way of thinking, however, continues to be noticeably reproduced in modern theological work. It is remarkable how close Cullmann's arguments are to those of the earlier Basel scholar, reproducing for the Greek words often almost the same points as were made by Orelli for the Hebrew. Cullmann's main thesis, that 'eternity' is not other than time, but a continuum or totality of time, is precisely that of Orelli; his argument from the plural in αἰῶνες is exactly that of Orelli from the plural in *'olamim*. Above all, the use of a lexical structure formed from the various words related to time is a method basic alike to Orelli and to Cullmann, although Cullmann has not retained the overwhelmingly etymological method of Orelli. Other lines closely parallel to Orelli's thought appear in other modern theological investigators who have discussed the subject of time.

The interweaving of philosophical assumptions and linguistic history, which is the essence of Orelli's method, probably fell somewhat into the background during the positivist philology of the later nineteenth century. In theological work on biblical language, however, similar methods enjoyed a certain revival along with the general theological revival of the 1930s and after,

and found most complete expression in the Kittel dictionary and
the modern tendencies of 'biblical theology'. In some ways the
more modern treatments show a degeneration compared with
that of Orelli. Thus Orelli was a good observer of linguistic
usage, and his fault was not that he did not notice usage but that
it tended to lose its full value once it had been integrated into
his dominant etymological scheme; but he often noticed how
actual usage failed to fit his general case. One cannot imagine
that Orelli would have simply passed unnoticed the evidence for
the sense of καιρός in Hellenistic Greek, after it had lain registered for decades in the standard dictionaries.

Another important difference must also be mentioned. The
modern theological school of interpretation of biblical language
tends to play off very strongly the Hebraic against the Hellenic
ways of thinking. Only in certain regards can support for this
be brought from Orelli. While we have seen that he contrasts
Hebrew with Greek and Arabic for the relatively large number
of its expressions for eternity (p. 105), and also points out the
lack of a mythological hypostatization of time in Hebrew thinking (pp. 106-110), many of the detailed points of linguistic fact
in Hebrew are not contrasted with, but illustrated by, points
from Indo-European languages. And other points made for
Hebrew thought are considered to be true of human nature
generally. Thus on p. 102 he maintains that 'the concept of an
extra-temporal being is foreign to Hebrew thought'; but just a
little later he says that such a being is foreign 'to human ways of
conceiving things'. Orelli's point in concentrating on Semitic
languages is not so much that their way of representing time is
unique, but that they show more clearly what can be seen also
but less clearly in Indo-European, the development in parallel
of knowledge and of vocabulary. The notion of a circular movement of time, which for the later Cullmann is Hellenic and un-
biblical, is amply evidenced for Semitic (including Hebrew) by
Orelli. When he sums up (on p. 63) his understanding of time
as something first expressed only in relation to particular times,
he says, 'it is in general not time, but *a* time, which is first appre-
hended by the mind and first named in language'. It is not clear
that 'the mind' here is only the Hebrew mind; rather the position
seems to be that the Hebrew language by its structure makes

more easily clear what is generally true for the human mind. Again, having proved that eternity in certain Semitic words is not something outside or above time but a continuum into which time runs, he claims this to be an illustration of Leibnitz's assertion that 'l'idée du temps et celle de l'éternité viennent d'une même source, car nous pouvons ajouter dans notre esprit certaines longueurs de durée les unes aux autres aussi souvent qu'il nous plaît' (p. 100). Leibnitz's assertion was certainly not intended to be valid only for the Semitic mind.

It is true that at times Orelli seems to claim that a special place is occupied by Semitic or Hebrew. Thus he points out the supposed absence of past, present and future reference in the tense system (p. 63 n.). Again, he says that words for time which regard it as 'a form similar to space and encompassing and enclosing all things' (this is the way in which he interprets Greek χρόνος and Avestan *zrvan*) are very much more common in Indo-European than in Semitic. But there are even more points at which no such opposition is made by him. On the whole Orelli cannot be claimed as an adherent to the sharp and systematized opposition of Greek and biblical thought which is characteristic of modern biblical theology.

There remains the assertion of Orelli that Hebrew has no word to express 'time as a quite general notion', an assertion which in one form or another appears frequently in the literature of modern biblical theology.[1] Orelli's own admission that this assertion was not true of usage without exceptions has received less attention. We must now consider how far the assertion is in any case a true one.

To begin with, we must reiterate that the question cannot usefully be identified with the question whether Hebrew has a word for 'time as an abstraction'. We have already seen that Greek χρόνος, the word most commonly cited as a contrast to Hebrew *'et*, does not mean 'time as an abstraction'. It can, however, be used in a type of statement about time which, whether we call it 'abstract' or not, might perhaps be reasonably taken to refer to 'time as a quite general notion', or 'time as a general entity'. Thus a philosophic usage like Aristotle's

---

[1] Marsh, *Fulness*, p. 20; *Word Book*, 'Time', p. 258; cf. Eichrodt in *ThZ* xii (1956) 107.

ὁ χρόνος ἀριθμός ἐστι κινήσεως κατὰ τὸ πρότερον καὶ ὕστερον
  (*Phys.* 220 a),

or a reflective dramatic remark like Sophocles'

ἅπανθ' ὁ μακρὸς κἀναρίθμητος χρόνος
φύει τ' ἄδηλα καὶ φανέντα κρύπτεται
  (*Aj.* 646),

could be said to be thoughts about time in its general and universal character. I repeat however that the word χρόνος here
used is also often used in classical Greek (like the Hebrew *'et*)
for particular times with reference to the events going on within
them. It is interesting to remark that in all this discussion, in
which so much seems to turn upon the presence or absence of
words for 'time in the abstract' or 'time as a quite general notion',
no one seems to have quoted an actual word in any language
which always or univocally means this.

We cannot therefore proceed usefully by labelling words in
general as 'abstract' or 'general', nor can we work, as has so often
been done in theological interpretation, by deducing from their
etymology an 'emphasis' which they must have had. Rather we
should consider a kind of syntactical context in which a word
commonly used for the time of some event may be used in a
more general sense in reflections about the nature of time. The
passages quoted above from Aristotle and Sophocles are examples
of such contexts. So also would be in English such passages as

'Time, like an ever-flowing stream,
bears all her sons away'
  or
'Time in the Bible is a straight line'
  or
'Eternity is not something different from time'.

It may be a matter of opinion whether 'time' in these contexts
bears a 'different meaning' from 'time' when used only in contexts related to particular events, or whether we should say that
a syntactical context of this reflective type simply uses in another
relation the same meaning of the word as has been used of particular times. At any rate the appearance of a Hebrew word for
'time' in a context of this reflective type should constitute *prima*

*facie* grounds for supposing that this word could be used of time in general. The non-appearance of the word in such contexts does not however prove the reverse; for it may only mean that such contexts are not found in Hebrew, that is to say, that no such reflection about 'time' in general was in fact entered upon within the extant texts, and does not thereby prove that words existing for 'times' in particular senses could not be used in such contexts, or that the non-existence of a word that could be so used is a reason why such contexts are not found.

It may at once be admitted that reflective contexts of this kind are not to be found in the main body of biblical Hebrew literature, namely in the legendary, historical and liturgical texts. But these texts are not characterized by reflection in general at all, except over God and his acts and the relations between him and mankind. In the wisdom literature on the other hand, with its urbane, educational and international atmosphere and its lack of express attention to the central traditions of Yahwism, we may well expect to find some contexts of the type we have defined.

At least one passage is at least probably so to be understood. At Qoh. 9.11 we hear that 'the race is not to the swift, nor the battle to the mighty, nor bread to the wise, nor riches to the discriminating, nor favour to the knowing, but time and chance befall them all' (*ki-'et wa-pega' yiqre 'et-kullam*). The most natural interpretation is that time in general is characterized by uncertainty and by vicissitudes. Accident may make talent vain.[1] The sense of *'et* in this context is surely not far removed from that of χρόνος in the Sophoclean context quoted above.

It must be added that this is only something to be expected in relation to the general philosophy of Qohelet. The stultifying and frustrating effect of time and change upon human effort is one of the main themes of this writer.

It is rather curious therefore that the great passage in 3.1-9, where 'there is a time for everything under the sun, a time for birth and a time for death' and so on, has been taken as an example of 'realistic time' in Marsh's sense.[2] While it may well be true that in detail the list gives 'times known in terms of their

[1] Cf. Galling, *Der Prediger Salomo* (HBAT, 1940), p. 83: 'Alles Geschehen unterliegt der Widerfahrnis der Zeit'. Also R. Gordis, *Koheleth—the Man and his World* (1955), p. 178, but cf. p. 297. See also BDB and KB s.v. *'et*.

[2] Marsh, *Fulness*, p. 21; his statement that Qohelet 'means that we can recognize

content', it is equally true and clear that the purpose of the whole
is to emphasize the frustrating effect of time on human life and
labour, whether because God has appointed the events beforehand or for some other reason.[1] The general trend of the passage
is surely not the rather complacent one that there is a time and
a place for everything; its conclusion is the despondent one
*ma-yitron ha-ʿosé ba-ʾašer huʾ ʿamel*, 'what is the advantage of one
who acts in all his toil?'. The uncertainties of time, and the
common experience that events only bring us back after much
toil to the point from which we started, are themes of this writer
in many other passages.

This interpretation is confirmed in a curious negative way by
the judgement of Wheeler Robinson. He represents the school
of thought which I have been criticizing, maintaining that the
etymology of *ʿet* 'suggests what the actual usage confirms, that
the Hebrew mind conceives time in the concrete, in its filled
content, and not as an abstract idea'.[2] But in his judgement about
Qohelet, to whose consciousness of time he devoted an appendix,
he at once declares the thought of this writer to be 'un-Hebraic',
and says that 'his time-consciousness, therefore, is useful as a
check on that of the Old Testament in general, by its very un-
likeness to this'.[3]

Now this judgement in itself involves several difficulties, to
some of which we shall later return. But whether we call the
thought of Qohelet Hebraic or not, and granting all possible
allowance to possible foreign elements in his thought and even
in his language, one can hardly say that the text of this writer is
not an example of Hebrew language, and one which must
generally have been intelligible to his contemporaries. In particular, his use of *ʿet* at 9.11 can hardly be denied value as evidence
for possible and intelligible Hebrew usage of this word. But my
suggestion that this is a case of the word in a reflective context
in which it may be taken to mean time in general is indirectly
confirmed by Wheeler Robinson's excepting of Qohelet from his
statement about the Hebrew conception of time.

those times by what takes place in them' can at the most be regarded as a very
indirect aspect of Qohelet's intention. See also *Word Book*, p. 258.
[1] See the review of possible interpretations of the list in Gordis, op. cit., p. 218.
[2] H. Wheeler Robinson, *Inspiration and Revelation in the Old Testament* (Oxford,
1946), p. 109.     [3] Ibid., p. 121.

Another case, although one only indirectly quotable, is also from the later wisdom literature, in Sir. 18.26: ἀπὸ πρωίθεν ἕως ἑσπέρας μεταβάλλει καιρός, καὶ πάντα ἐστὶν ταχινὰ ἔναντι κυρίου. The meaning here is surely not that 'opportunity' or 'decisive moment' keeps changing all through the day, but that time passes swiftly from morning to night. The sense is 'time' in general. The Hebrew text is not available here, but it is most probable that Sirach used *'et*, which is quite common in his book, and possible that he used *zᵉman*, which also occurs.

It is true that both Qohelet and Sirach are very late representatives of the wisdom literature, and it seems likely that reflection about time in general appears only late in this tradition, or at any rate in what we have of it. It may on the other hand be an open question how far the general thought of Qohelet about the frustration of human activity by temporal realities may go back into antiquity. It would be a reasonable and conservative suggestion to suppose that some such theme had been handled from early in the post-exilic period in certain limited circles. The appearance of the actual word *'et* in a reflective syntactic context like Qoh. 9.11, even if this was the first case to occur, is therefore probably hardly a drastic innovation or adjustment, and the use of *'et* for time in general may therefore without undue rashness be pronounced a possibility within post-exilic Hebrew.

To these wisdom passages can be added an absolutely certain case from Qumran, where in a hymn describing the government of the day and night and seasons by the heavenly luminaries we have the phrase *tmyd bkwl mwldy 't* 'continually in all the generations of time' (the translation of Burrows).[1]

This brings us to another point which is of importance for the subject as a whole. In the comparisons of Greek and biblical thinking it is common to adduce and compare the lexical material of New Testament and classical Greek and classical Hebrew. But the vocabulary of Jesus and of Paul, whether they used Hebrew or Aramaic or both, was not that of classical Hebrew. In addition to the familiar classical *'et* 'time', they would almost certainly use *zᵉman*, which appears only a few times in late passages of the Old Testament; in Aramaic they may also have used *'iddan*, which appears in the well-known Danielic language

---

[1] 1QH 12.7-8.

about 'times and seasons'. But in particular we have to reckon
with the probability of an extensive familiarity with *qeṣ* in the
sense of 'period, time'. This is of course frequent in the Qumran
documents; it appears also in Rabbinic usage, and especially
interesting for us is the use there for the eschatological periods
of world history.[1]

The word *qeṣ* also presents one or two pieces of evidence for
arguments in which we have already been engaged. Firstly, the
sense of 'time in general' is at least a possible, and I should think
a probable, sense in the Hebrew text of Sir. 43.6, where the
moon is said to be *mmšlt qṣ w'wt 'wlm*; this may be reasonably
translated as 'government of time and sign of eternity'. The
Greek however with its

$$\mathrm{\mathring{a}\nu\mathring{a}\delta\epsilon\iota\xi\iota\nu\ \chi\rho\acute{o}\nu\omega\nu\ \kappa\alpha\mathbf{\mathring{\iota}}\ \sigma\eta\mu\epsilon\hat{\iota}o\nu\ \alpha\mathring{\iota}\hat{\omega}\nu os}$$

perhaps took it to mean particular times defined by the movements of the moon. Secondly, a number of cases of *qeṣ* have been
translated by καιρός in the LXX. Some of these may not be
quite correct translations, and for example the other meaning of
the Hebrew word, i.e. 'end', may well be more likely at Gen. 6.13.
But the fact that the Hebrew word is translated by καιρός at a
time when the usage of this Hebrew word was certainly mainly
of periods and epochs, as at Qumran, is further proof that it is
wrong to try to understand the NT καιρός as 'decisive moment'
wherever possible (as *TWNT* does), or to argue with Cullmann
that 'the characteristic thing about the kairos is that it refers to
a *moment* defined by its content'. It suggests indeed that at the
critical theological passages of Acts 1.7, Eph. 1.10, to quote only
two, if we are going to be more precise than 'time' we should
perhaps go not in the direction of 'moment' but in the direction
of 'epoch' or 'period'.

The evidence of late Hebrew usage thus gives us one or two
possible or probable cases of words for 'time in general', and also
some useful evidence for the interpretation of New Testament
words.

There remains another Hebrew word which might be mentioned in connection with time in general, but which I mention
with some reserve because it appears in a rather limited series of

[1] See for instance Jastrow s.v.

contexts. I refer to the well-known word *yom*, which of course commonly means 'day'. This word has been enlisted by Marsh in his series of terms for realistic time, and usages like 'the Day of the Lord' have been emphasized: 'days, like times, are known and identified by their contents, and one content—the activity of God—is emphasized as fundamental'.[1] But a more general sense of 'time' appears in certain contexts in Hebrew usage from very early times, and is registered in the standard dictionaries. This sense appears mainly in the plural; to give one example, *yamim* at Num. 9.22 is certainly 'an indefinitely long period' (BDB). BDB in fact devote a paragraph to usage meaning 'time', and Orelli remarks that the use of *yom* in the plural with the sense of 'time' has been noticed since Kimchi's time.[2] I do not press this example, however.

There is another aspect in which words for distinct periods of time have been used in this discussion. It has been argued that periods like 'day' or 'hour' were not precisely measured chronological quantities, and that therefore in many cases they should be understood as representing 'realistic time', with the emphasis on actions occurring and not on temporal quantity. Thus Pedersen argues[3] that:

'Nor, when the day is divided, is it according to an abstract measure dividing it into equal lengths; the decisive factors are the different peculiarities of the sun which leave their different impress upon everything. . . . The colourless idea of "hour", measuring time in a purely quantitative way, is far from the old Israelitic conception.'

Quoting this passage, Wheeler Robinson asserts that:

'Classical Hebrew had no word for "hour", and the term *ša'a* assimilated from the Aramaic in post-biblical Hebrew denoted a brief space of time, rather than an arithmetical fraction of the day. The Hebrew divisions of the day were of a quite general kind. . . . The point of interest in these rough-and-ready divisions of the day and night is that they are characterized by their content. . . .'[4]

---

[1] *Fulness*, p. 26 f.; cf. *Word Book*, p. 260 f.   [2] Orelli, p. 52 n.
[3] *Israel* i-ii. 489.   [4] Op. cit., p. 107 f.

Now it is certainly true that the word *ša'a* 'hour' does not appear in biblical Hebrew, although it appears in biblical Aramaic. One must nevertheless be very cautious in pronouncing on these grounds that it did not exist. We here reach the limits which are set to linguistic assertions about a dead language which has left only a very restricted body of literature. There are numerous objects and realities of Hebrew life for which we do not know the words by which they were designated. The non-occurrence of the word in the literature is not proof that it was unknown. We may therefore proceed to consider on other grounds what the probabilities are. Here another point must be remembered. If we suppose that *ša'a* 'hour' was an existing and intelligible word in classical Hebrew, we do not assert that precise chronological measurement was used or was possible. Only an approximate standard of accuracy, and one that might vary seasonally or with other circumstances, need be envisaged.

Now the description of short periods of time by reckoning up a number of 'hours' was already familiar to Rib-Addi, regent of Gubla, in the fourteenth century, when he writes (admittedly in Accadian): *inuma ištu* 10 *šeti kašadiya ana beruta uširti mariya ana ekalli rubi* 'when ten hours after my arrival in Beruta I had sent my son to the sublime court',[1] where *šeti* 'hours' is considered to be possibly a Canaanitism by Knudtzon.[2] It is surely difficult to suppose in view of this that at the court (say) of Solomon, a king who had certainly not a lower level of material achievement than Rib-Addi and who was far from isolated from non-Israelite culture, some such phrase as *'eśer ša'ot* 'ten hours' would never have been used or would have been unintelligible. The interpretation of the famous passage about Hezekiah and the turning back of the sun at II Kings 20.8 ff. has been much disputed, but recent opinion seems to favour the view that some kind of sun clock was involved.[3] Such an instrument would almost certainly have precise divisions, as Egyptian examples (including those found in Palestine) have, and it is surely extremely probable that

[1] J. A. Knudtzon, *Die El-Amarna-Tafeln*, 138. 75 ff.
[2] Knudtzon, p. 1521.
[3] See J. A. Montgomery and H. S. Gehman, *The Books of Kings* (1951), p. 508, and literature cited there. For a sun clock found in Gezer and belonging to the 13th century B.C., see W. R. Sloley, 'Primitive Methods of Measuring Time' in *JEA* xvii (1931) 166-78, and in particular p. 173. See also the article 'Sun-dial' by H. Spencer Jones in the *Chambers' Encyclopaedia* (1950), xiii. 286 f.

Hebrew had a word which could be used for the corresponding period of time.

One must therefore regard as doubtful the suggestion that the Mishnaic word *ša'a* 'hour' is a loanword from Aramaic. It is more probable that late usage here simply retains a word which was known in classical times but which does not appear in the texts. Of the Aramaic word itself something should also be said. It is true that, in the Old Testament Aramaic texts where it occurs, it is probably not used of a temporal measurement exactly, but rather in a general sense of 'time', giving phrases like 'at that time'. This however does not show that the word was not also used for a measured period. In the New Testament we have many cases where ὥρα is used in a rather general sense, as in πρὸς ὥραν 'for a time', but it is also well known for a measurement of time, as in 'the ninth hour' and so on. The usage in the Aramaic Daniel therefore by no means certainly shows that at that time the word was not also used for an arithmetical fraction of the day. Even if it should, however, be true that the Aramaic word usually means 'moment' or the like (cf. *ša'ta'* in the Targum for Hebrew *rega'* 'moment'), then there is all the less reason to suppose that the post-biblical Hebrew usage, where the sense of an arithmetical fraction of the day is well known, depends upon the Aramaic.

We conclude, then, that an examination of vocabulary does not support attempts to depict a special Israelite way of thinking about time for which the quantitative senses of words like 'hour' or 'day' are relatively unimportant. It is not unlikely that some of the reasoning which leads to such a special way of thinking is an application to Israel of what is supposed to be a 'primitive' human apprehension of time.[1] In this very matter of time measurement, however, it is probable that with the influence of neighbouring civilizations the Israelites were far removed from the 'primitive'.

[1] Thus what is commonly said about the concrete and non-arithmetical nature of the Hebrew appreciation of time is very much the same as is said about 'primitive' indications of time by M. P. Nilsson, *Primitive Time-Reckoning* (Lund, 1920), p. 355 and *passim*. Cf. G. Pidoux, 'A propos de la notion biblique du temps', *RThPh* ii (1952) 120-5, especially p. 121. But there is extreme difficulty in applying all this material taken from 'primitive' cultures to the culture of at least post-exilic Israel, with its carefully cultivated chronological sense.

# V

## VOCABULARY STOCKS FOR 'TIME' AND THEIR TRANSLATION

### A. THE COMPARISON OF LEXICAL STOCKS

THE procedures which we have been criticizing make two, probably unconscious, assumptions:

(a) They assume that the layout of the lexical stock of a language forms a pattern symmorphous with the patterns or structure of thought supposedly common among the speakers of the language concerned.

(b) They assume as a consequence that the variations between languages in respect of this layout correspond to differences in the thought or mental pattern typical of the peoples or groups using these languages.

Thus by noting the existence of a pair of contrasting words for 'time', whether καιρός and χρόνος or καιρός and αἰών, they read off a structure of the mental understanding of time in the New Testament. Again, noting that Hebrew has one word *baśar* 'flesh', while Greek has two words, σῶμα 'body' and σάρξ 'flesh', they may assume that this difference can be explained *only* by the existence in the Greek mind of distinctions such as that between form and matter, while the Hebrews made no such distinctions, so that one word was enough for them.[1]

It has already been shown how seriously the linguistic evidence has been distorted *in fact* where these procedures have been used on words for time. It remains to consider in a more general and theoretical way the nature of the distribution of lexical stocks, so as to criticize these procedures rather *in principle*.

Two preliminary points may be made here. Firstly, full value should be given to the fact that lexical stocks are to a considerable extent inherited. If an area of a lexical stock is surveyed diachronically, that is, over a considerable period of historical change, all kinds of alterations will be observed: words going out of use, new words appearing, new senses becoming usual, and so on. The factors affecting these changes will be various

[1] Robinson, *The Body*, p. 12 f. For fuller criticism see *Semantics*, p. 34-38.

and complex. Some will be connected with general cultural change, while others will have more purely linguistic causes, such as phonological changes. Only a small area of such change will be traceable to the prevalence in changed degree of general philosophical distinctions such as (to take Robinson's example) that of matter and form. The historical perspective therefore should make us cautious in correlating the layout of lexical stocks too closely with mental-philosophical systems, because (*a*) it will be extremely unlikely that more than a small proportion of the historical changes, by which the lexical stock of any period has in the past been formed, can be traced to the influence of such systems, and (*b*) because in some cases the lexical items discussed will have existed long before the emergence of the mental-philosophical system correlated with it—such is the case for example with Greek σῶμα and σάρξ, which existed in the Greek vocabulary long before the distinction of matter and form was formulated, and which in any case were not correlated with that distinction, even after it had been formed, by Aristotle, the most prominent user of that distinction.

Secondly, we should remember that it is possible for philosophers and others to enunciate a distinction between two words, a distinction by which they express an important difference believed to exist in actual things or in mental attitudes. It might then be possible to speak of the deliberate formation of lexical structures which are intended to represent, or to coincide with, actual structures in the external world, or in ways of looking at it. This is done when we 'make a distinction'. Examples might be the distinction of 'form' and 'matter' in Aristotle, or of 'phenomena' and 'noumena' in Kant, or between '*geschichtlich*' and '*historisch*' in certain German theology. Similar distinctions could doubtless be quoted from the biblical area. But it is important to realize that the lexical stock of a language as a whole is not organized in this way, and the fact that a pair of words, with some distinction between them, exists in some area does not demonstrate that a distinction of this philosophical type is the origin of the existence of the varying terms. On the contrary, the making of a distinction in the philosophical sense is rather the systematization within a certain type of thinking of certain differences which already previously existed within the lexical stock. It may be that the prevalence of attempts to demonstrate a lexical structure in biblical language is a result of an approach to that language made by scholars whose primary interests lie in philosophical-theological problems rather than in linguistic methods.[1] Whether this be so or not, it is clear that one cannot take any pair or group of words from biblical language, such as καιρός and χρόνος, and, because certain differences of sense exist between them in certain contexts, assume that they are always or indeed ever used to point or express a systematic distinction of the philosophical-theological type. Where such systematic distinctions occur, we should remember, evidence of this fact will probably be clearly presented. There is not a sentence in the Greek Bible in which καιρός and χρόνος are juxtaposed in order to express a systematic contrast of the philosophical-theological type.

These then are two general points, one about the historical or diachronic study of lexical stocks, the other about the relation of philosophical distinctions to differences between lexical items generally. In the rest of this chapter the approach will deal with synchronic description of lexical stocks. The stocks of certain languages will be compared as they stand at certain periods, for example, in classical Hebrew, or New Testament Greek, or modern English. For this approach all the emphasis lies on the stock as it stands at the time under discussion, and the past derivation or history of items in it is not relevant.

We may therefore begin to orientate ourselves to the study of lexical stocks by making a very general survey of how lexical items in a number of different languages are disposed in relation to the English word 'time'. We might agree that this word has different 'senses'; we may often sense, without much further analysis, that in one context it means a rather different sort of thing from what it means in another context. We often test this by word-substitutions. Thus there is part of the range of 'time' over which we may say that 'occasion' does not contrast appreciably. Thus 'three times in the year' is commonly very similar

---

[1] See also *Semantics*, pp. 275-8 etc. In general it may be suggested that some of those forms of modern theology which have most completely repudiated 'philosophy' in favour of biblical authority have often merely transmuted the biblical language systems, or their lexical stocks, into a kind of 'logic' or 'philosophy'. To do this at all means allowing linguistic fact to be forcibly dominated by theological-philosophical patterns. This is, of course, bad theology and bad philosophy as well as bad biblical linguistics.

to 'on three occasions in the year'; the contrast here would probably be felt to be stylistic for the most part, the latter phrase being rather less common and more stilted. One would not go far wrong in saying that in phrases like this 'time' and 'occasion' do not contrast. In a dictionary of English it would not be surprising to find 'occasion' registered for such cases as one of the meanings of 'time', and in a bilingual dictionary we might find a foreign word used in such phrases registered as 'time, occasion'. On the other hand there is part of the range of 'time' where 'period' does not contrast very much, and few people would sense a contrast when for 'in the time of the Judges' we substituted 'in the period of the Judges'. But we could not substitute 'in the occasion of the Judges' without a drastic change of meaning. Thus while 'occasion' and 'period' will each in certain contexts present little or no contrast with 'time', they will in very many contexts, or in all, present great contrast with one another.

Similar methods can be used if we translate words from one language to another, or if we want to compare the lexical stock of one language with that of another. If we start from English 'time', we know that for many instances we shall say French *temps*, German *Zeit*, and Hebrew *'et*; but for certain instances where the English sense overlaps with that of 'occasion', as in 'three times in the year', we shall say French *fois*, German *Mal* and Hebrew *pa'am*. Some languages will at one point afford more possible alternatives than another. Thus for 'at that time' in late Hebrew one could probably use either *'et* or *zᵉman*; in the LXX, as we have seen, one can use καιρός or χρόνος indifferently in such contexts, but the former is more common. 'Three times in the year' can in some languages be done with an adverb like Greek τρίς 'thrice', Latin *ter*; but this alternative is not necessarily available, as for instance in French. In the LXX some books say τρίς (Kingdoms in fact, also Daniel for Aramaic) for the phrase which in the Pentateuch is several times τρεῖς καιροὺς (τοῦ ἐνιαυτοῦ). For 'seven times', on the other hand, the Pentateuch says ἑπτάκις very frequently. The identical Hebrew phrase in four contexts fairly close together is translated by Jerome as *ter in anno*, *tribus temporibus anni*, *ter in anno*, and *tribus vicibus per annum* (Ex. 23.17; 34.23, 24; Deut. 16.16, and compare the *tribus vicibus per singulos annos* for a synonymous Hebrew phrase at Ex. 23.14).

Some words which mean 'time' will also have another quite different sense, that is to say, will correspond to a quite other unit in the English lexis. Thus *temps* in some contexts means English 'weather', as does καιρός in modern Greek; in classical Greek καιρός, which in some contexts means 'opportunity', in others means 'right measure', while in biblical Greek the same word means 'opportunity' in certain contexts but in others means 'time' in general and is interchangeable with χρόνος. Hebrew *pa'am* also means 'sole of the foot'.

All this is nothing new, and is indeed normal experience in linguistic communication. The above is no attempt to make an exhaustive analysis of lexical stocks relating in some way to time, and only picks out a few obvious features. One could continue indefinitely describing the conditions in which a word for 'time' in one language overlaps or does not overlap with a word in another. For this there is a linguistic reason. The vocabulary of a language is not a structure or a system in the way in which for example the grammar or the phonology is. All these overlaps and non-coincidences, of which only a small proportion has been cited above, cannot possibly be explained as the reflection of different sets of temporal conceptions held by the speakers of the different languages. Surely no one will undertake to expound a feature of the French character which demands a concept of time which will also insist on including the weather, and which will be easily contrasted with an English concept of time which excludes the weather but demands the expression by the one word 'time' of what is distinguished by *temps* and *fois* in French. Yet arguments of exactly this kind have been common in the modern theological interpretation of Hebrew and Greek thought and language. The main reasons why this has been done I have suggested elsewhere[1]; they are the isolation of the comparison of Hebrew and Greek from general comparative linguistic method, and the absence of any real description of these languages as a whole, with the use only of such phenomena as appear to fit with contrasts already formed in the realm of philosophical-theological comparison.

The most important point then that emerges from our brief study of lexical stocks is that the comparison of these must be

[1] *Semantics*, p. 21 ff.

based, and can be based, on a method that is strictly linguistic, being founded largely on the consideration of word-substitutions in given syntactical environments. In other words we once again repudiate the treatment of a word as a 'concept' in the way which has been normal in modern biblical theology. Such a method normally leads naturally to an attempt to regard the location of the various items on the vocabulary grid of a particular language as reproducing the essential elements of a thought structure in the typical mind of the members of the speech group concerned. The attraction of this method is that it enables the layout of a group of words within the lexical stock to be made to serve in the establishment of some essential point in a philo-sophical-theological comparison. It very commonly does this however only at the expense of the serious inaccuracies which are stimulated by the essential characteristics of the method, and of which we have seen a glaring example in the case of καιρός and χρόνος as interpreted by Marsh and Robinson. The tests against which most 'concept' treatments of words in theological study fail are: firstly, the test by application to the words in their syntactical contexts, which is for instance immediately fatal to such an example as the interpretation of Hebrew *dabar* by Procksch, followed by Torrance.[1] The 'concept' method is really a method of de-syntacticization of the words. Secondly, the test by ex-tension to other words than the few where it might at least appear to work. Thus a reader might be willing to accept Cull-mann's αἰών concept which embraces both time and eternity, but will perhaps begin to doubt its validity if the same method leads (and why should it not lead?) to a French '*temps* concept' which includes both time and weather, and which is an essential element in the Frenchman's world outlook.

Some attention may be given at this point to the importance of translation. Just as, in dealing with one language only, the method of making the simple word-substitutions like replacing 'time' in various contexts by 'occasion' or 'period' is a great convenience in examining the lexical stock, so in dealing with several languages is the ability to replace 'time' by '*temps*' or '*fois*' as the case may be. In this case the trying out of such simple substitutions involves also the tentative trying out of others for

[1] See below, p. 112 f.

the other words of the syntactical environment. But, granting this additional complication, the use of simple translation equivalences is an important instrument in lexical investigation.

I have already elsewhere called attention to the disparagement of translation which is a rather natural result of the 'concept' treatment of words now fashionable in theology. This may take the simple form of Kögel's contempt for dictionaries which just furnish a translation equivalence and a registration of occurrences, and which are thus good enough only for schoolboys.[1] It can take the more refined form of believing that Hebrew is untranslatable, or translatable only by virtue of a special divine act or intervention.[2]

Now it would no doubt be unwise to exaggerate the positive importance of translation as an interpretative method. On the other hand we shall see that it furnishes a valuable and perhaps essential check in investigations of lexical stock, and that neglect of it can have disastrous results. In more general terms we may say that, unfortunate as it may be to regard simple translation as a central aim of linguistic study, especially in the case of word-for-word translation, nevertheless the knowledge of a substantial number of translation equivalences is a basic element in the knowledge of a foreign language, just as the progressive learning of more and more such equivalences is essential to the learning of a language.

The disparagement of translation equivalences has been especially marked in the appreciation of Hebrew in recent theology, and has its origin in the belief that this language is characteristically heteromorphous in relation to Western languages, just as the Hebrew mind is supposed to be peculiar in relation to the Western, and most of all the Greek, mind. Thus in an article developing this point of view we find the statement that 'translations which represent a Hebrew word in all cases by one English word are absurd'.[3] But this is surely a most unlikely and ill-considered statement. How is it absurd that *keleb* should

---

[1] *Semantics*, p. 243 f.

[2] C. Tresmontant, *Études de métaphysique biblique* (Paris, 1955), p. 253 ff. See *Semantics*, p. 265 n.

[3] A. R. McAllaster, 'Hebrew Language and Israelite Faith', *Interpretation* xiv (1960) 421-432, especially p. 425. This article presents in a condensed and convenient form the main opinions which I have criticized in my *Semantics*.

be in all cases translated by 'dog' or *šata* in all cases by 'drink'? A very considerable number of such cases could be quoted where a classical Hebrew word can be rendered by the same English word throughout the Bible, and where careful and responsible versions like the AV and RV have in fact done so. There are of course also words where things are not so simple, and where quite a large number of equivalences have to be given in order to give a picture in English of the sense even in quite common examples. Is it certain however that the number of such words in classical Hebrew is bigger than it is in Latin or French? To stick to our examples of words for time, after all, surely nearly all cases of *'et* in biblical Hebrew can be translated by 'time', which is more than one can say of *temps*, where many examples will mean 'weather'.

This is perhaps enough for a preliminary justification of a translation procedure as an important and responsible procedure in a lexical enquiry. It is a specially valuable procedure in modern theological interpretations of language, because the 'concept' treatment of words often produces statements which claim to indicate the 'real meaning' of words, but which are totally impossible when tested as translation equivalences in actual syntactical contexts.

A good example is that of *dabar* as interpreted by Procksch.[1] He says that this word means the hinterground of a thing. 'Every thing has its *dabar*, that is, its hinterground and meaning.' Now whether or not Procksch's treatment has some interest for Israelite psychology in a very general way, there is a very strong reason why one must take it as a very doubtful suggestion for the semantics of the word *dabar*: there is no extant context of this word in biblical Hebrew in which it can in fact be translated by 'hinterground', and conversely, there is surely no imaginable English sentence containing the word 'hinterground' which could be translated into biblical Hebrew, or any kind of Hebrew, by using *dabar* to represent this English word. While it is not necessary for us at this point to say that statements of meaning which cannot be related to translation equivalences are necessarily wrong, it would at least be advisable to distinguish them clearly

[1] *TWNT* iv. 90. Cf. my analysis of a similar treatment in *Semantics*, p. 129 ff.; also my *JSS* article on 'Hypostatization'.

as two quite different sorts of thing. In fact it is clear that a great many among such alleged 'real meanings' are plays on etymologies in the interest of some theological bias.

There is another reason why translation procedures perform a valuable service. They provide the necessary antidote to the de-syntacticization of words in the 'concept' method. Translation depends upon the recognition and understanding of the syntactical environment in which words occur. Thus, to cite Procksch again, the assertion that 'every thing has its *dabar*, that is, its hinterground and meaning' is made invalid by the fact that the syntactic context to which this could apply, a sentence like 'all things have their *dabar*' or 'the *dabar* of this thing', does not exist. The statement in the explanation of usage of *dabar* that 'every thing has its *dabar*' thus implies a wholly fictitious syntactical context, and thus is valueless for the explication of biblical Hebrew *dabar*. In fact it is clear that Procksch must have produced this wholly baseless interpretation through a burst of philosophical inspiration stimulated by the etymology.

Indeed the protection of the interpreter against false arguments from etymologies is one of the great services which translation checks provide. While in the concept method a word is seen as a 'concept' and placed within a frame of related 'concepts', the patterns of the frame reflecting the realities of theological truth, translation takes an actual communication in one language and translates it into an actual communication in another. Etymological considerations, which have so often been employed in order to assign words to their proper place in the theological 'concept' framework (as with *dabar* above), may very commonly not have been a consideration in the actual communication at all, and in such a case the task of translation will help to keep etymological considerations within their proper bounds.

The use of the concept method, on the other hand, can often be identified by the appearance of bogus translations, or sometimes of the assertion that the word is untranslatable.[1] When etymological considerations have been used to 'bring out the inner meaning' of a word, one is commonly offered etymologizing

---

[1] For an example of the latter, cf. Barth's treatment of σπλαγχνίζομαι in *KD* iii/2. 252; E.T., p. 211; see *Semantics*, p. 156 f. Of course there may well be many words which are genuinely untranslatable within certain limits.

or literalizing translations which are supposed to reproduce this. Here is an example of an argument on the subject of time, which depends on mistranslation:

> 'The Hebrews conceived of time quite differently from us. To the Hebrews time was identical with its substance. Time was identical with the development of the very events that occurred in it. For example, the OT offers us such phrases as: "Time is rain" (Ezra 10.13)'.[1]

The Hebrew here is *ha-'et g<sup>e</sup>šamim*. The meaning is quite properly indicated by a translation like the RSV, which I quote at some greater length in order to show the context: 'But the people are many, and it is a time of heavy rain; we cannot stand in the open'. Knight's translation 'Time is rain' makes no sense whatever in the context, and merely ignores the treatment of the passage in the standard translations and reference books.[2] It is a literalizing translation produced purely for its effect in the current argument. But even taken with all literalness, it is obvious that the determination by the article here gives the sense 'the time' or 'this time' and not 'time'. The construction generally is quite a common one in Hebrew, and is registered in the standard works.[3]

One may cite also the attempt of Torrance to relate time to the nature of the Church by means of a lexical structure based on the word στοιχεῖον.[4] This is an example of a structure built upon the multiplicity of senses in a word.[5] This treatment involves:

(a) etymologization: 'the root meaning of στοιχεῖον, which it never lost, is succession';

(b) hypostatization: 'this does not mean that the Church must have nothing to do with στοιχεῖον';

(c) the use of literalizing and etymologizing translations along with recognized or correct ones: the meaning of τῷ κανόνι

---

[1] G. A. F. Knight, *A Christian Theology of the Old Testament* (London, 1959), p. 314.

[2] See for example BDB and KB; GK § 141 d; Brockelmann, *Hebräische Syntax*, 14 b ε; W. Rudolph, *Esra und Nehemia*, p. 94.

[3] Incidentally, the argument is an example of how Knight, though arguing for the extreme 'concreteness' of Hebrew diction, has to make what is in fact fairly 'concrete' into a kind of abstraction in order to make it fit his arguments.

[4] 'Atonement and the Oneness of the Church', *SJT* vii (1954) 262 ff.; *Royal Priesthood* (Edinburgh, 1955), p. 52 ff.

[5] Cf. the survey above, p. 16f.

τούτῳ στοιχήσουσι is given as 'proceed in time according to a new canon', although normal renderings like 'element' are also used for στοιχεῖον;

(d) inconsistency: for it is said that στοιχεῖον 'never lost its root meaning', i.e. 'succession', and yet well-known senses like those for physical elements, letters of the alphabet, rudiments of philosophy, and the shadow in a kind of sundial, are quoted;

(e) far-fetched connections: the sense of στοιχεῖον for the sundial shadow is supposed to be relevant for Col. 2.17, 'which things are a shadow of things to come';

(f) the much-debated question, which of the many senses of στοιχεῖον was intended by Paul, becomes an idle one, for apparently Paul intended them all, plus the 'root meaning' in addition to all extant senses, and on top of these an etymologizing play on στοιχεῖν. This example is cited here partly to give another instance of the fascination of lexical structures, but mainly to indicate how a responsible translation of the Pauline passages would necessarily be fatal to such an interpretation, and how, conversely, the propounding of such an interpretation naturally produces examples of bogus translation.

All this goes to emphasize the importance of translation as an element in responsible handling of biblical material, and to justify the importance which has been laid upon translating ability in our traditional language teaching. I repeat that translating is important because it means the taking of a passage in one language as a communication, and rendering this into a communication in another language. It involves the leaving aside of etymological information, and of senses which the words have elsewhere than in the communications being handled.[1] I do not say that translation is the ideal way of approaching or handling biblical language; but I do say that it forms a valuable check and corrective of such interpretations as these. It may be said plainly, if with regret, that interpretations which flourish Greek and Hebrew words are not therefore more accurate than others. Some of the interpretations based on Hebrew which have been criticized by the writer in his *Semantics* contain, in spite of their superstitious awe

---

[1] The central importance of *the communication intended* as a basis for exegesis is rightly seen and emphasized in quite another connection by C. D. Morrison, *The Powers That Be* (London, 1960), p. 63 ff.

of the Hebrew language, numerous mistakes which would have been impossible if some reasonable heed had been given to a reasonable translation like the AV; or, to put it the other way, the subjects studied could have been more satisfactorily treated on the basis of the AV than has been done by the artificial and etymologizing treatments supposedly based on the Hebrew.[1]

We may now summarize the conclusions for this section. Firstly, the distribution of lexical stocks is normally haphazard between one language and another, and between one language and the mental habits typical of the speakers of that language. The words for a subject like time do not form a structure from which can be read off the essential structure of thought about time within the members of a linguistic group. The attempt to do so only produces serious distortions of recognized semantic data, as we have seen in the cases of Marsh and Cullmann. But even if they had taken all the material into consideration and avoided their more obvious errors, their basic plan would still be a wrong one. Secondly, care in the recognition of translation equivalences is an essential element in a proper linguistic examination of lexical stocks. To the linguistic emphasis of translation which has been made here we may add a theological point: the possibility of a sufficiently accurate translation of the Bible is a matter of the utmost concern for the life of the Church.

### B. WORDS FOR TIME IN THE ANCIENT BIBLICAL LANGUAGES

*Biblical Hebrew*

With these points in mind we may now proceed to give a brief survey of the more important words for time in the Bible and of the questions involved in their translation from one language to another. This is by no means a complete survey, but only an attempt to gather together the principal points, some of which have been made in detail in the more controversial part of my argument.

In biblical Hebrew '*et* means 'time' and no other word is of comparable importance for the discussion of time. A very large proportion of occurrences are translated by 'time' in English

---

[1] For examples see *Semantics*, pp. 129 ff. and 161 ff.

versions. For cases like 'seven times', that is, the sense of 'occasion', Hebrew more commonly uses other words, especially *pa'am*, but *'et* is used in this sense, 'many times', in the latish passage Neh. 9.28.[1] A common setting is in a phrase like *ba-'et ha-hi'* 'at that time', giving the temporal setting of an incident in relation to others.[2] Another is the statement that 'it is an evil time' or the like, where the description is probably of conditions prevailing over a period (e.g. Amos 5.13 or Pss. 9.10; 37.39). 'At every time' or 'always' is *bᵉ-kol-'et* quite frequently. Another frequent setting is 'the time of' something, that is, the time proper for it or appointed for it. I have already argued that 'time' in general reflection about time is a context that appears in some wisdom passages. Finally, the plural seems to come close to 'experiences, fortunes' (BDB) in I Chron. 29.30, *ha-'ittim ᵃšer 'abᵉru 'alaw*, and following this 'fortunes' or 'destinies' may be good translations at Isa. 33.6; Ps. 31.16[3]; if so, these are examples of transition somewhat out of the strictly temporal sense.

Mention should also be made of *mo'ed* 'appointment', which can be used for an appointed time as well as for the place or fact of meeting, and which comes to be used especially of sacred seasons fixed by the calendar or the heavenly bodies.

The word *'olam* refers to the remotest time, or means 'perpetuity'. Of the past we may hear that something is *me-'olam* 'from of old, from earliest times', and for the future something may be *lᵉ-'olam* or *'ad-'olam* 'for ever'. Attached to objects now existing it means something like 'perpetuity'; thus an *'ebed 'olam* is a 'slave in perpetuity'. A common context is in relation to the nature of God and of institutions created by him, which are *lᵉ-'olam* or 'perpetual'. There are few, or perhaps no, examples of a sense 'eternity' in a free syntactical context such as 'eternity is like so and so' or 'I am thinking about eternity'.[4]

[1] The text has been questioned here on the basis of the Greek. If the MT is wrong here, then this case does not occur in biblical Hebrew. But many authorities accept the MT; see Rudolph's commentary ad loc., for example.

[2] Marsh's attempt to use this phrase as a sort of technical reference to the classic time of God's activity is an extreme case of over-interpretation of a simple phrase; *Fulness*, pp. 56-65; *Word Book*, p. 259 f.

[3] H.-J. Kraus, *Psalmen* (1960), p. 250 and p. 81, mentions Albright's interpretation of *'ittot* as 'portent', but appears not to follow it in his treatment; his actual translation at Ps. 31.16 is 'my future', which is perhaps adding rather too much to the literal sense of 'my times' or 'my destinies'.

[4] See Jenni's discussion of the difficult passage Qoh. 3.11 in *ZATW* lxv (1953)

## *Alterations in later Hebrew*

Several important changes have appeared by the time of post-biblical documents, such as the Qumran literature on the one hand and Mishnaic sources on the other. The difference between Qumran usage and Mishnaic usage is considerable; on the other hand in each the usage lies to a considerable extent within certain rather specialized literary types. For this reason it seemed better to present here some alterations *vis-à-vis* the usage of biblical Hebrew, rather than to try to work out another overall picture for post-biblical Hebrew. It must be considered probable that both Qumran usage and Mishnaic usage may be relevant for the background of the New Testament.

At Qumran, *'et* continues to be frequent for 'time'. But for periods of time in the divine periodization of historical epochs *qeṣ* has become very frequent indeed. In some cases *qeṣ* may mean a moment rather than an extended period, as in 'the time of visitation' at CD 19.10 f. For 'appointed time' *mo'ed* continues to be used, especially in theological passages about the allotted times fixed by God. At Qumran *'olam* is very frequent, and common contexts are such as: *klt 'wlm* 'eternal destruction'. *'š 'wlmym* 'eternal fire', *lkwl qyṣy 'wlmym* 'for all the ages of eternity'. As in the Old Testament, *'olam* is used predominantly in phrase contexts such as construct groups and prepositional phrases.

In Rabbinic usage the place of *'et* as the common word for 'time' has been extensively taken by *z<sup>e</sup>man*, which in the Mishna is much more frequent. As at Qumran, *qeṣ* is found used for times and periods related to the divine purpose, and sometimes ex-

24 ff.; but Galling's interpretation, which he follows, involves a change of text and other difficulties. Gordis accepts the sense of 'world' here and interprets it as 'the love of the world'—see op. cit., p. 221 f. Perhaps I might offer, with much diffidence, another possible interpretation, which would work from the well-known sense 'perpetuity' and also avoids any change of text. I would suggest translating by 'He has done everything well in its time; also he has set perpetuity in their heart (i.e. in the heart of man)—yet so that man cannot find out the action which God does from the beginning to the end'. The reference to perpetuity would mean the consciousness of memory, the awareness of past events. The predicament of man is that he has this awareness, and yet cannot work out the total purpose of God. This avoids the artificiality of the old rendering 'eternity', understood as the consciousness of another world beyond this one, and of the sense 'the world', given by Gordis, which requires the rather difficult amplification as 'the love of the world'. On my interpretation no unusual meaning would be given to the word, but only a rather unusual context for a well-known meaning. It requires no alteration of text, and is consonant with the Greek version.

plicitly of the time of the Messiah's coming (e.g. bMeg. 3 a). For *'olam* on the other hand, besides the continuation of OT usage, we have in Rabbinic usage the sense of 'age', i.e. a period as opposed to another, and so particularly in the formulae like *ha-'olam haz-ze* 'this age' and *ha-'olam hab-ba'* 'the coming age'. Moreover, we have the sense of 'world' in addition to the temporal senses.

## The Septuagint

For LXX usage a good deal has been said already. We may however add a reminder that the language of the LXX occupies a peculiar place; it is a translation register, which varied in important respects from the actual speech of the Hellenistic Jews, and which at certain points and in certain respects can be understood only in relation to the translation techniques being used in the various sections. Moreover, these techniques are usually not uniformly applied, even within a single block of the translation; and, finally, the LXX includes books which were not translated from Hebrew at all. These considerations, which may apply to some degree to all the ancient translations, are probably specially important for the LXX, and have to be borne in mind in the making or use of any general lexical surveys.

In the translation of *'et* καιρός is much more frequent than χρόνος, but it is clear that they are not infrequently interchangeable. In phrases of the elapse of time, with quantity marked by an adjective, χρόνος is more common. The classical use of καιρός, where we would translate by 'opportunity', is not common, but occurs, e.g. probably in I Macc. 9.7 and very probably in Sir. 12.16. On the other hand a lifetime can also be designated by καιρός, as in Wisd. 2.5 σκιᾶς γὰρ πάροδος ὁ καιρὸς ἡμῶν 'our lifetime is the passing of a shadow' (with variant reading βίος). It can also mean 'the conditions' (Sir. 29.5), it can be used of 'time' as a general entity in free contexts (Sir. 18.26), and as used in translation of Hebr. *mo'adim* it can mean 'holy seasons, festivals' as in Bar. 1.14 ἐν ἡμέρᾳ ἑορτῆς καὶ ἐν ἡμέραις καιροῦ. Ὥρα also is used of time, especially in translating *'et* in phrases with meanings like 'at this time tomorrow', and also in general senses with 'time', such as those which we might translate as 'all the time', 'at the right time' and 'in the time of'.

Αἰών is very frequent in the LXX, but an overwhelming majority of cases are in the bound phrases like εἰς τὸν αἰῶνα 'for ever'. Those phrases which in Hebrew are done with noun in the construct, like *bᵉrit 'olam* 'everlasting covenant', are in very many cases done in the LXX with the adjective αἰώνιος, as in διαθήκη αἰώνιος, κατάσχεσις αἰώνιος and νόμιμον αἰώνιον. The rendering with the noun αἰών is however found, and is actually the commoner construction in Sirach, where we have σημεῖον αἰῶνος 'eternal sign', εὐφροσύνη αἰῶνος 'eternal joy', and so on. The case of the 'permanent slave' at Deut. 15.17 is done in Greek with καὶ ἔσται σοι οἰκέτης εἰς τὸν αἰῶνα.

The sense 'eternity' in a free syntactical context is uncommon; a case will be found at 4 Macc. 17.18 τὸν μακάριον βιοῦσιν αἰῶνα 'they live through blessed eternity', and Qoh. 3.11 in Greek may well have been understood in this way.

The sense 'world' is very probable at Wisd. 13.9 and certain at Wisd. 18.4: δι' ὧν ἤμελλεν τὸ ἄφθαρτον νόμου φῶς τῷ αἰῶνι δίδοσθαι.

The old classical sense of 'man's life' seems to be meant at Ps. 89(90).8, in view of the general context there, although the rendering itself has been caused by the reading of the Hebrew as *'olamenu*.

The sense of a particular 'age' or epoch cannot be certainly evidenced from the LXX for αἰών; it is, however, possible that this sense, which is known from the NT, was already current, and that it affected the understanding of some passages, e.g. the one or two cases of πάντων τῶν αἰώνων and the like, Ps. 144 (145).13 and I Esd. 4.40. The reading πατὴρ τοῦ μέλλοντος αἰῶνος at Isa. 9.6(5), which is a clear case of this sense, is not part of the original Septuagint.

## New Testament Greek

When we come to the Greek New Testament we are again in a living language system, and not in a translation register valid only for the text of a sacred and defined canon. This is so even after allowance for the influence of translations from Aramaic in the Gospels. The relation of the Greek of the Gospels to its Semitic originals is rather different from that of the Septuagint to its Hebrew original. Still more outside the Gospels, as in

Paul's letters, does the language exemplify the things that were actually said.

Both καιρός and χρόνος are used commonly where we would translate 'time'; as in the LXX, the former is commoner, but the proportion of χρόνος is rather higher. In many contexts the two words are interchangeable, apart from the stylistic preference for καιρός. For the lapse of time, with an adjective of quantity, χρόνος is usual; for cases like 'the time for figs', and for 'opportunity', καιρός is used. The strong eschatological expectation, and the sense of fulfilment of past eschatological promises, produces frequent contexts like 'the time is coming', in which ὁ καιρός is usual. The sense of 'season', both for natural and sacred seasons, appears a few times for καιρός. In two passages both χρόνος and καιρός appear together, probably with no appreciable difference in meaning.

Ὥρα is also often used, not only of the division of the day, but also of 'time' more generally. One common context is 'the time is coming when . . .'; another is 'from that time on', with the impression of 'immediately thereafter'; another is of a short period of time, as in πρὸς ὥραν 'for a short time'. We also have the context 'it is time' or 'it is high time' to do something, and 'the time of' someone, that is, his moment, when he has some decisive part to play. The frequency of this word is strikingly higher than in the LXX.

Αἰών, as in the LXX, continues to appear in the bound phrases of the type of ἀπ' αἰῶνος 'from of old' and εἰς τὸν αἰῶνα 'for ever'. But the use with the sense 'age', and nearly always in contexts like 'this age' and 'the coming age', is very common. The sense 'world' as in 'he created the world' is also certain, although there are cases where one may hesitate between 'this world' and 'this age'. Cases like 'perpetual fire' and 'eternal redemption' are done with the adjective αἰώνιος, as is more usual in the LXX, and not with the Semiticizing construction with the genitive. No case of a free context meaning 'eternity' is found.

We now turn to the translations of the Bible, in which we shall attend only to the New Testament versions. Here again, as with the LXX, we have to consider in each case how far the idiom of the original and the translation techniques affect the language of the version and its value as a specimen of actual usage in the speech of the Christian community using this version.

## Latin

In the Latin Vulgate, the chief point to remark is that all cases of καιρός and of χρόνος are alike translated by *tempus*, with very few exceptions indeed.[1] The exceptions include one case where μικρὸν χρόνον is translated by *modicum*, although in this case *tempus* is usually present; and the two cases where both χρόνος and καιρός appear together.

These two cases deserve some special attention. Jerome writes *tempora vel momenta* at Acts 1.7, and *de temporibus et momentis* at I Thess. 5.1. If there is something special in καιρός here for Jerome, it lies not in the rightness or the decisiveness of the time, but in the brevity of its duration; for his use of *momentum* compare its use to render στιγμή at Luke 4.5. In both NT examples the first word of the two is χρόνος. At Wisd. 8.8 the order is the opposite; the Latin gives *temporum et saeculorum*.[2] In all cases, then, the first in order of the two words was rendered by *tempus*, the normal equivalent for both; but when *tempus* had been thus pre-empted for the first word, different shades of meaning had to be expressed when equivalents *other than tempus* had to be given for each of the two Greek words.

In no case does Jerome feel it necessary to bring out a peculiar sense of καιρός by translating, say, with *opportunitas*, the word which he used for εὐκαιρία. Jerome's rendering *opportuno tempore* at Acts 24.26 (25) depends on the ἐπιτήδειος in the reading καιρῷ δὲ ἐπιτηδείῳ known from some Greek MSS, where most have καιρὸν δὲ μεταλαβών. It would seem unlikely that the Latin versions depend on a mere unsensitivity to the variety of the Greek words. Jerome must certainly have known the classical distinction between καιρός and χρόνος; we have seen that Augustine

---

[1] This brief survey is based on the Vulgate as it left Jerome's hands. It would of course be relevant to consider the effect of the Old Latin on his usage; but this would be complicated, and a brief survey by the writer suggested that it would not materially affect the immediate subject.

[2] This is of course a pre-Jerome rendering. It is also interesting that at Dan. 2.21 the LXX καιροὺς καὶ χρόνους was rendered simply by *tempora* in the Old Latin (text in E. Ranke, *Fragmenta versionis antehieronymianae* (Vienna, 1868), pp. 48 and 110), and that Jerome has used *tempora et aetates* in rendering from the Aramaic.

both knew it and used it for an important point of interpretation. Jerome knew and used a Stoic definition of καιρός as χρόνος κατορθώσεως which is also quoted by Philo,[1] a definition intended to bring out the opposition with the sense of χρόνος as 'time, period'. But Jerome not only avoided pressing the New Testament examples into this sense by the use of translations like *opportunitas*, he also interpreted cases of καιρός in the general sense of 'time'. Thus for the 'redimentes tempus, quoniam dies mali sunt' of Eph. 5.16, where the Greek has τὸν καιρόν, Jerome's exposition seems clearly to envisage that the intention is to refer to the employment of extended periods of time, and makes no reference to 'decisive moments' or 'opportunities' or the like. Thus he says: 'quando in bono opere tempus consumimus, emimus illud, et proprium facimus quod malitia hominum venditum fuerat'; and again: 'redimentes tempus . . . dies malos in bonos vertimus, et facimus illos non praesentis saeculi, sed futuri'.[2] We need not insist that this is the right interpretation of the passage. We only maintain that Jerome, although he certainly knew of the classical sense in which καιρός was distinct from χρόνος, did not emphasize this sense in his Latin translations; and that this may well be a result of the attention to biblical usage which was forced upon him by his constant work of translation and interpretation. He knew, after all, that καιρός in Daniel sometimes means 'year'.[3]

The use of *hora* in the Latin version coincides with that of the Greek.

Where the Greek has αἰών, however, some variation appears. The general equivalent for αἰών is *saeculum*, which appears in the bound phrases like *a saeculo* 'from of old', *in saecula saeculorum* 'for ever and ever', and in the cases meaning a particular 'age', as in *hoc saeculum* and *saeculum venturum*, and in those meaning (probably) 'world', as in *per quem fecit et saecula*. The Latin however uses some variety in many of these contexts; thus for the same Greek phrases we sometimes have *in aeternum* or *in dies aeternitatis* or *in sempiternum*. For the adjective αἰώνιος the usual rendering is

---

[1] Jerome in *PL* xxv. 1393: unde et Stoici quibus curae est verba singula definire, tempus dixerunt esse correctionis sive efficientiae; quod significantius Graece dicitur χρόνον εἶναι κατορθώσεως. Cf. Philo *Op. Mund.* 59.

[2] *PL* xxvi. 559f. This follows a Greek interpretation.

[3] *PL* xxv. 534: 'tempus annum significat'.

*aeternus, saecularis* appearing only in the phrase *ante tempora saecularia* (for πρὸ χρόνων αἰωνίων), and *caelestis* and *sempiternus* also appearing. An interesting point is that *saeculum* is used not only to translate αἰών but also to translate κόσμος, though not frequently; but κοσμικός is *saecularis* in both its two occurrences. The two chief points here, then, are the use for αἰών of a word which also is sometimes used for κόσμος 'world', and the predominant use for 'eternal' or 'perpetual' of a word which has no etymological connection with *saeculum*.

## Syriac

In Syriac one must be careful because of the different versions of the Gospels and of other books, and because of the different histories of the versions, and in particular the different history of the Epistles from that of the Gospels. The following survey is based on the Peshitta and the Old Syriac Gospels. Study of the Apocalypse involves the Philoxenian.

Within these translations the word for 'time' is in overwhelming preponderance *ẓabna'*, which is used for the vast majority of cases of both καιρός and χρόνος. Orelli states that it is 'almost the only word for time in Syriac'.[1] There is however also the word *'edana'*, also meaning 'time', and it appears very sporadically in Syriac biblical texts. Occasionally it appears in a context like 'the hour is late' or 'the time of incense', where the Greek has ὥρα; so Matt. 14.15 pesh; John 16.4 pesh; Luke 1.10 sin pesh; 14.17 sin pesh; Acts 3.1. In the Apocalypse *ẓabna'* is used, except at 12.14 in an OT quotation, where *'edana'* is used following the OT text.[2]

There is, then, no effort to use the two Syriac words in order to reproduce the difference between the Greek καιρός and χρόνος.[3] For the two places where the words are used closely together, at I Thess. 5.1 the Syriac says *ẓabne' w'edane'*, but at the other,

[1] Orelli, p. 57.
[2] It is interesting that in the Syriac version of the Aramaic parts of the OT *ẓabna'* was considerably preferred, and was used several times where the original had *'iddan*, e.g. Dan. 2.9, 21. Sometimes *ša'ta'* also was used for *'iddan*, e.g. Dan. 3.5. But in Dan. 4 and 7 *'edana'* was used, hence the quotation in the NT. Contrast here the Jewish Targums, which stick to *'iddan* with much consistency as a rendering for *'et*.
[3] One may compare the Palestinian Syriac version, which used *qyrws* for καιρός and *'šwn* for χρόνος with great consistency.

Acts 1.7, it says *zabna' 'aw zabne'* 'the time or the times'.[1] For 'hour', Greek ὥρα, the Syriac uses *ša'ta'* with very sporadic exceptions. For 'always', Greek πάντοτε, it uses the phrase *bkul zban*, but sporadically, as at John 7.6, *bkul 'edan*. Compounds of καιρός, like εὔκαιρος and πρόσκαιρος, are also sometimes done by prepositional phrases with *zabna'*. On two occasions καιρός is translated by the word *qirsa'*, itself a loanword from Greek καιρός. The preference for it here may probably be ascribed to a desire to avoid the juxtaposition of *zabna'* with the verb *zabnin* 'buying', which would produce a strange paronomasia (Eph. 5.16; Col. 4.5). Generally speaking, then, the Syriac works with one word for 'time', and very sporadic use of others.

If the Latin version translated αἰών with a word which was also sometimes used for the Greek κόσμος 'world', the Syriac lies farther in the same direction, for *'alma'* is the *standard* rendering for both αἰών and κόσμος, with only occasional exceptions. For the adjective αἰώνιος, on the other hand, Syriac uses no adjective form, and renders the Greek by phrases containing *'alma'*. The adjective *'almanaya'* means 'worldly' and translates κοσμικός (Heb. 9.1). At one place where αἰών and κόσμος come together, at Eph. 2.2, in κατὰ τὸν αἰῶνα τοῦ κόσμου τούτου, the Peshitto writes *'aik 'almayuteh d'alma' hana'* 'like the worldliness of this world'.

## Coptic

In Coptic, as in Syriac, care must be taken to distinguish the usages of various versions and dialects. The following is restricted to the Sahidic New Testament:

Here, as with Syriac *zabna'*, the word for time in the vast majority of cases is *yoiš*, translating both καιρός and χρόνος. Other words were available and are very sporadically used, such as *sey* 'time, season', which appears at Mark 11.13, and *te* 'time of maturity or readiness', which appears at Heb. 11.15 in the phrase 'they would have made opportunity'. Furthermore, the Greek words could also be used as Coptic lexical units, and χρόνος in fact so appears at Acts 1.7; I Thess. 5.1 (the two passages

---

[1] Compare Wisd. 8.8, where the two Syriac words are presented in the same order as at I Thess. 5.1, but where, as has already been seen, the Greek words are in the reverse order, with καιρός coming first.

where it appears along with καιρός in Greek), while καιρός as a loanword appears sporadically, as in Mark 12.2. There was no attempt to use two different words consistently to represent καιρός and χρόνος respectively. That the loanword χρόνος at Acts 1.7 and I Thess. 5.1 has not been brought in in order to bring out any special force of this word in the Greek original can be seen because (a) its place in the Coptic is that of the καιρός of the original, and (b) in Acts 1.6 χρόνος itself is translated by the usual word, ųoiš. As in other versions, the juxtaposition of χρόνος and καιρός has led to a translation which was not normal for the words taken separately.

The Sahidic also shows an interesting contrast in the rendering of αἰών. The phrases like εἰς τὸν αἰῶνα 'for ever' are rendered by Coptic phrases with *eneh*, for example *ša eneh*, 'for ever'. In those cases, however, where αἰών means a particular age, as in 'this age' or 'the age to come', the Sahidic writes the Greek word αἰών. (Here there is an important difference from the Bohairic, which uses *eneh* in this type also.)

Have we gained anything by this brief description of the main aspects of the lexical stocks relating to time in the languages of the older biblical versions? At least we have gathered some information together, which may serve as a counterbalance to the analysis of the arguments of others, which has formed so large a part of this book. What else has been achieved?

(a) We may claim to have shown here, in however inadequate a way, that work on lexical stocks both must and can proceed strictly on the basis of usage, and in particular that the picking out of odd words, or of particular senses of words, which appear to 'fit' with the general point of view of 'the Hebrews' or 'the Greeks', is both unnecessary and wrong.

(b) We have surely shown that it is impossible to understand or to explain the layout of the lexical stock of a language in respect of time (as probably of most things) as a 'reflection' of a general attitude to time supposedly held by the community speaking this language. The explanation of a lexical stock and its distribution as a 'reflection' of a *Weltanschauung* in this way may appear faintly plausible when applied to only one or two languages, such as Greek and Hebrew or Greek and Latin; but the staunchest will surely faint before the task of producing seven or eight *Weltanschauungen* to go with each of the speech groups here studied, to say nothing of the dozens of others one could add.

(c) We have surely shown that the description of a lexical stock within any area of usage at any time does not require any application of etymological methods—and naturally enough so, since the speakers themselves largely did not know, or did not bear in mind most of the time, the facts of etymology. In fact most appeals to etymology in modern biblical theology have been illegitimate, and have been used in order to produce an 'emphasis' in a word in particular contexts, just where usage alone can be a guide. Etymological study can of course be properly applied to the material we have assembled here, but it must be clear that it will not produce any evidence which can overrule usage, for its material is valid and legitimate only when understood as strictly historical material. The use of etymological material to form a kind of mental substratum which is supposed to underlie all usage, and from which various emphases and connections can be read off, is quite illegitimate. Historical study of lexical stocks will, however, rightly bring to our notice the extent to which such stocks are inherited, and thus give us another powerful argument against the idea that they 'reflect' or 'fit in with' some world-outlook of the period of the texts under discussion.

(d) While opinions may differ about how far each translation has succeeded in communicating the essentials of the text it was rendering, it would surely be hard to maintain that large and essential areas of the message communicated by the New Testament were quite lost in languages other than Greek because such languages did not reproduce the shape of the Greek lexical distribution. If there is such a thing as a 'New Testament view of time', can it be maintained that its outlines are wholly obscured for the reader of the Latin or the English Bible because these languages do not reproduce the supposed structure of the Greek καιρός and χρόνος? If it is so important that αἰών is a 'concept' which includes and stands above the distinction of time and eternity, then is the reality thus reflected wholly lost to the reader of the Syriac Bible, whose word included the world as well, or to the reader of the Sahidic Bible, who had different words for two different senses of the Greek αἰών? The impossibility of answering these questions in the affirmative leads us back to the faultiness of a method which tried in the first place to distil the biblical concept of time from the layout of the lexical stock of New Testament Greek.

We may state then that the accurate knowledge of the distribution of a lexical stock is necessary for the understanding of statements in the relative language (and my summaries here cannot claim to be quite accurate, since they are very brief—and yet lexicography can never be complete), but that one cannot attribute 'meaning' to this distribution, or regard it as a reflection or a simple product of an outlook on life. Theological meaning belongs to the *utterances*, in which elements of the lexical stock are syntactically combined with others.

# VI

## THE APPROACH TO THE PHILOSOPHICAL
## AND THEOLOGICAL QUESTIONS

THIS book does not propose to propound or to argue for any theory of time or of eternity, and all that is intended in this chapter is a reconsideration of how the questions of a deeper theological and philosophical kind may be posed, in so far as the posing of them depends on arguments from linguistic phenomena in the Bible. In so far as they may or should be approached from outside the Bible altogether they will not be discussed here, nor will it be asked whether or not it is legitimate within Christian theology that such an approach from outside strictly biblical territory should be made at all. It is clear however that a very large number of discussions in recent times, which have intended to reach a theological-philosophical position, have in fact attempted to work from biblical statements and from biblical linguistic phenomena. To clarify the problems involved we may begin by distinguishing a number of different kinds of statement that may be made. Each will be accompanied by examples (the citation of which does not imply my belief that they are true statements).

1. Statements about lexical phenomena, such as:
    'Hebrew has no word for abstract time', or
    'Hebrew cannot translate χρόνος', or
    'The same Greek word, αἰών, is used both for limited periods and for unlimited time or eternity'.
2. Statements about morphological and syntactic phenomena, such as:
    'The tense system of the Hebrew verb does not express distinctions of past, present and future'.
3. Statements about the meaning of biblical passages in which words for 'time', 'eternity', etc., occur, such as:
    'Qohelet says that there is a time for everything', or
    'The Gospels frequently speak of the coming of an expected age, which is often called "the coming age" '.
4. Statements about the general thought of biblical writers, based on passages in which such words occur, such as:

'For Qohelet time is recurrent and meaningless', or

'For the early Christians the passage of time is understood to be a movement towards an eschatological goal'.

5. Statements about the temporal implications of biblical statements which do not themselves contain words for 'time', 'eternity', etc., such as:

'The statement that the Word was made flesh implies a realistic view of time', or

'Time only began to be meaningful at that point in history when God brought his people out of Egypt'[1].

6. Statements about the general characteristics of the mind of a large group like 'the Greeks' or 'the Hebrews', such as:

'The Greek view of time was essentially cyclic', or

'To the Hebrews time was identical with its substance',[2] or

'To the Greeks time was unreal, and salvation consisted in an escape from time'.

In general the most obvious thing about the modern discussion in theology is that it has been assumed that statements from these different categories can simply be added together, or that a statement in one category can be used as evidence to prove an assertion in another. Marsh's work in particular is an example of the interweaving of statements of these different kinds, along with further philosophical-theological argumentation of kinds which have not been categorized above. The assumption that this is a right method is very common in modern theology.

One or two of the categories call for some small mention in particular. Category 2, that of statements about morphological and syntactic phenomena, is nearly always exemplified by claims that the tense system of the Hebrew verb, expressing aspect and not time, can be fitted into some theory of the Hebrew view of time. Boman, who has gone farthest with such claims, has been criticized by me elsewhere, and I shall not repeat these criticisms here.[3] Though many theologians have tried to connect the tense system with a Hebrew way of thought about time, in spite of or in ignorance of Bauer's powerful arguments as far back as 1910,[4] and though Knight, typically of this approach, continues to call

---

[1] Knight, *Christian Theology of the Old Testament*, p. 102.
[2] Ibid., p. 314.    [3] See *Semantics*, pp. 72-82, etc.
[4] In *BAss* viii (1910) 1-53; see *Semantics*, p. 41 f.

the tense system an 'element in Hebrew thought',[1] few of them have really made anything out of this point, and, except for Boman, they have attached much less importance to it than to the lexical arguments. In fact we can no more deduce a Hebrew view of time from the verbal tense system of Hebrew than we could deduce the Platonic and Aristotelian philosophies of time from the tense system of Greek, if these philosophers had never written or if all trace of their writings had disappeared.

But, we may ask, is the same not true of statements in category 1? Is it not true that we can no more deduce a biblical view of time from the lexical stock of Hebrew or of New Testament Greek than we could deduce the Platonic and Aristotelian philosophies of time from the lexical stock of classical Greek, if these philosophers had never written or if all trace of their writings had disappeared?

Category 1 has in fact supplied the more important arguments to which we have given attention. It is worth while to summarize the reasons why this category of statements has been so prominent.

(a) The theory, which has been opposed by me in this study, that the words for an entity form a structure or pattern from which the structure of thought about that entity can be read off, formed the explicit basis of the first modern study of the matter, by Orelli.

(b) An attention to lexical study on these general lines, and with the belief that biblical language contained patterns symmorphous with the outlines of biblical-Christian thought, has long been current in biblical theology; it was fostered by Cremer, but received a great stimulus with the planning and publication of *TWNT*, and has now become extremely fashionable.

(c) The modern movement in biblical theology has included a turning away from, and eventually in most cases an almost complete ignorance of, modern general linguistic methods which might have suggested other ways of handling the lexical phenomena.

(d) Perhaps more than any other factor, however, is one still to be mentioned—the very serious shortage within the Bible of the kind of *actual statement* about 'time' or 'eternity' which could

---

[1] Knight, op. cit., p. 292.

form a sufficient basis for a Christian philosophical-theological view of time. It is the lack of actual statements about what time is like, more than anything else, that has forced exegetes into trying to get a view of time out of the *words* themselves. This is so important that it deserves a more extended discussion.

We have seen in the last chapter that although words which we may rightly translate as 'time' and 'eternity' in certain contexts occur plentifully in the Bible, the syntactical contexts in which they occur are limited in a number of ways. Thus the following could be said to be reasonable representations of typical biblical contexts:

> 'God has promised times of restoration'
> 'Jesus said that his time had not yet come'
> 'Christ is alive unto all eternity'
> 'The Gospel claims that the coming age has already arrived'.

But another type of context is conspicuously rare, such as:

> 'Time is the same thing as eternity'
> 'Paul teaches that eternity is not timelessness'
> 'Time is the field of God's action'
> 'Time is known by its content'
> 'God created time'
> 'There is a time, other than our time, which is God's time'.

It is immediately recognizable that this latter group have characteristics which hardly any passages in the Bible display. They are actual statements setting out to say something about time, that 'time is so and so' or 'eternity is this and that'. It is equally obvious that, though this group is hardly represented in the Bible at all, it is precisely the kind of statement which the biblical theologians writing about time have tried to formulate. The biblical passages about which statements of category 3 may be made are themselves not representative of this second group of sentences. The result is inevitably a concentration on category 1 statements for linguistic evidence; and the assumption that these will certainly 'fit' with favourable statements in category 6 produces the distortion of lexical facts and the general confusion of argument throughout, which are characteristic of recent theological discussion of the matter.

The first necessity, then, for advance in the philosophical-theological discussion of time is the making of adequate distinctions about the category of the statements made, whether about lexical phenomena, about particular biblical passages, or about Greek or Hebrew thought in general, and the realization that statements made in one category cannot be validly transferred into another, or correlated simply with statements made in another, without careful methodological consideration and justification.

The books we have discussed are clear examples of this need. Marsh, as we have said, shows throughout a close interweaving of lexical, of exegetical, and of general philosophical-theological statements. But the attempt to combine these different elements spoils them all. The philosophical ability of the work could have been so much more fruitful if the author had only left καιρός and χρόνος alone. Conversely, however, it was the unjustified general belief that the layout of a group of words closely or exactly reflects the structure of thought which in the first place drew Marsh into stating a distinction between them which fitted his primary philosophical distinction, but which glaringly departed from known and recorded linguistic usage.

Cullmann on the other hand professes to reject philosophy and philosophical argument in his approach to time. Thus 'we must make ourselves completely free from all philosophical concepts of time and eternity if we want to understand the use of aion by the early Christians'.[1] One may suspect that to Cullmann 'philosophy' means idealist philosophy; but this need hardly concern us here. What is more important is that Cullmann, while repudiating philosophy and with it all theology which uses philosophical insights, and while insisting that the concrete biblical material is the sole foundation of truth, by his lexical method has hypostatized and conceptualized the words, assimilating them to mental concepts and concealing their actual linguistic function in usage. By de-syntacticization of them he has produced a structure out of his words which fits precisely with his view of time, although the words are nowhere used in the texts to *state* this view or anything similar to it. In fact the hypostatization of vocabulary items is one of the principal ways

---

[1] Cullmann, p. 41; E.T., p. 48.

in which modern biblical theology has forced dogmatic and philosophical schemata upon biblical material, while at the same time
professing on the one hand to be non-philosophical and non-dog-
matic, and on the other to be exact and scientific linguistically.[1]
Thus it is only apparently that Cullmann's structure of time is biblical and non-philosophical. In this respect Marsh is more realistic,
and freely admits the philosophical aspects of his method.

The faults in lexical treatment which this approach produces
have already been studied in detail. It remains to look at the
general statements about lexical method made by Marsh and
Cullmann. Marsh writes:

> 'The words that the biblical writers used will repay our
> study in giving some indication as to their ideas of time,
> eternity and history. Words are symbols, and symbols, more-
> over, which disclose something of the realities they express.
> To study them usefully is to pass beyond the range of philology.
> The derivation of many Hebrew words is uncertain and
> obscure, and these lie behind the Greek usages in the New
> Testament. It is more important to note how words are used,
> to discover in the light of what convictions about time, eternity
> and history certain terms gain currency'.[2]

To this one must say:

*a.* What reason have we to assume that words 'disclose something of the realities they express'? What does 'dirt' reveal about
dirt, or 'horse' about the nature of horses? One way in which
words can thus be made to reveal or 'disclose' something about
the realities they designate is by etymologizing, the method
which was explicitly embraced by Orelli and which has spread
disastrously through modern theological treatments of language.
Though Marsh himself here rightly shows a reserve towards
etymology, it is the conception of words which he himself
embraces which has encouraged etymologizing treatments.

*b.* Why must we pass beyond the range of 'philology'? Though
Marsh's reason is not quite clear, his statement naturally reads

---

[1] For a discussion of linguistic hypostatization see my forthcoming article in the
*Journal of Semitic Studies.*

[2] *Fulness,* p. 11. For a criticism of the phrase 'lies behind', commonly used of
the relation of Hebrew words to Greek, see my letter to *The Times Literary Supple-
ment,* March 31st, 1961, p. 201.

as if philology were primarily interested in etymology and in historical linguistic investigation, and that by investigating actual usage one goes out of the area of philology altogether. But this would be an extremely narrow and unusual understanding of the sense of 'philology'. Even in that period (now past) in which the emphasis of linguistic study was very much more heavily on historical and etymological research than it is now, it would have been unheard-of to maintain that it did not also include the study of usage and its relation to current trends of thought. It is thus very difficult to see any meaningful sense of the word 'philology' by which the study of word usage would be excluded. On the other hand one must remark that just in those 'theological' treatments of biblical language in recent times, which might perhaps suppose themselves to be going 'beyond the range of philology', etymological considerations have very frequently been seriously over-valued against the evidence of actual usage, as many examples in my *Semantics* show. Though Marsh himself avoids any false emphasis on etymology, others have not done so, and it must be admitted that modern theological treatments of language have often produced manifestly exaggerated etymological emphases which would have been impossible if any reasonable contact had been maintained with modern non-theological linguistic studies.

*c.* As we have seen, Marsh himself deprecates etymologizing treatments, and proposes to work from the history of usage. But we have already seen that his treatment of καιρός is in manifold respects remote from a correct picture of usage. The basic fault in his method can be seen in the delightful vagueness of 'in the light of' in the last sentence of the quotation above. What is the actual relation between general convictions about a reality and the linguistic change of use of the words designating it? The vagueness of Marsh's statement invites one to assume that the two go neatly hand in hand. We know that the Christians held strong general convictions about the critical decisiveness of certain events in time and history. We also know that at certain times and in certain contexts καιρός is distinguished from χρόνος by meaning 'right time' or 'critical time'. Now this word is common in the New Testament. Is it not natural then to suppose that it there means 'critical time' or 'realistic time', and that its frequency in this sense should be 'seen in the light of' this general Christian conviction? If we suppose that it is natural to understand word usage 'in the light of' general convictions of this kind, are we not inviting what has in fact happened in this case, a failure to observe that the usage is quite different from what we might expect 'in the light of' this general conviction? And when we see that in fact καιρός in biblical Greek, as in extra-biblical Hellenistic Greek, frequently or indeed preponderantly does *not* mean 'right time' or 'critical time', how do we begin to understand this 'in the light of' Christian convictions about time and history?

Cullmann also, and more clearly than Marsh, shows his intention to form a lexical structure which will be symmorphous with the reality of time and eternity. He refers to certain dictionaries such as that of Kittel, which has tried to bring out the theological meaning of the words. But, he goes on, 'certain necessary limits are set to theological understanding where a *single* word is handled separately. Therefore it is our wish to show in a connected survey how this terminology itself expresses the peculiar nature of early Christian thought about time'.[1]

Thus from the lexicography of the words taken separately Cullmann goes on not to the exegesis of the syntactical contexts in which they in fact occur, but to a new kind of context, a framework in which the words have their places and which coincides with the outlines of the New Testament view of time. Thus at the end of this chapter, after his lexical structure is complete, he can state that 'the terminology of the New Testament teaches us that' such and such is the Christian view of time. We are to be taught, then, not by actual sentences or communications spoken or written by the men of the New Testament, but by a systematization of the conceptualized words.

It need hardly be pointed out that such a position, quite apart from its linguistic validity, must be extremely precarious theologically. That some kind of authoritative value in teaching, similar in some respect to the authority of the statements and communications of the Bible, should be conceded to lexical structures of the terminology, must be very dubious doctrine within any currently accepted view of biblical authority.

The first necessity for future argument, then, is a lexical method

[1] Cullmann, p. 32; E.T., p. 38f.

which is freed from the pseudo-linguistic methods I have criticized, and a philosophical-theological method which does not confuse itself by trying to draw conclusions directly from lexical material; and above all a general awareness that nothing is to be gained by an indiscriminate mingling of the two methods. It is worthy of note that we are not asking for something entirely new in this. Theological treatments of time which make no attempt to take a lexical structure as their basis already exist, and exist among theologians who would certainly claim to be observing biblical authority in the utmost degree.[1]

We now turn to one or two problems in the general description of mental attitudes (statements of category 6).

In much of the recent discussion it is confidently asserted that 'the Greeks' held what is called 'a cyclic view of time'; and for Cullmann in particular it forms the basic distinctiveness of the biblical view of time as against the Greek that in the former time is a straight line and in the latter it is a circle.[2] Marsh agrees with Cullmann in regarding the 'Greek view' as a cyclical one and in rejecting it, although he has difficulties with Cullmann's straight line.[3] Now even if we accept as true a certain amount in the contrast here set forth, there are a number of difficulties in it and a number of ways in which it has surely been greatly exaggerated. The following criticisms suggest themselves: (a) 'the Greeks' did not always in fact hold a cyclic view of time; (b) 'the Hebrews' in certain cases and in certain respects did entertain a cyclic view of time; (c) that which 'the Greeks' regarded as cyclic (if they did) was not the same thing as that which 'the Hebrews' regarded as a straight line (if they did).

We may first of all however mention briefly those accounts of the matter which try to deny that time had any importance for 'the Greeks' at all. Thus in 1922 von Dobschütz published an article[4] in which among other things he argued that, though the

---

[1] Thus the long treatment of time by Barth in *KD* iii/2. 524-780, E.T., pp. 437-640, appears not to work from a lexical structure of biblical words, whatever other criticisms one may have of it. So also E. Brunner, 'The Christian Understanding of Time', in *SJT* iv (1951) 1-12; Camfield in *SJT* iii (1950) 127-48; Dillistone in *SJT* vi (1953) 156-64.

[2] Cullmann, p. 44, and 43-52 generally; E.T., p. 51, and 51-60 generally.

[3] Marsh, *Fulness*, p. 175, 179.

[4] 'Zeit und Raum im Denken des Urchristentums', *JBL* xli (1922) 212-223.

Greek philosophers produced detailed theories about space, they did nothing of the same kind for time, and his point of view continues to be quoted, for example by Cullmann.[1] That von Dobschütz's argument is sheer nonsense is obvious in view of the lengthy discussion of time by Aristotle, to take only one example, and this was already remarked by Delling,[2] and even by Boman, although the latter follows von Dobschütz in his more general notion that Greek thought is dominated by the spatial and Hebrew thought by the temporal.[3] In favour of his thesis von Dobschütz pointed only to the fact that for the Hebrews the paradise was removed in time from the present while for the Greeks it lay in a far-away place, and connected this with a general statement from Windelband that the Greeks found no metaphysical interest in the course of time, saw no planned connection in historical development, and never thought that such development might be the area where the real essence of the world lay. For our purpose the article of von Dobschütz is of importance only to indicate the enormous generalizations about 'the Greeks' which have been acceptable coin in modern theological discussion. Most works have in some form held that Hebraic and Greek conceptions of time differed in that time was very much more meaningful to the former than to the latter.[4]

We come, then, to the opinion that 'the Greeks' held a cyclic view of time. We need not doubt that some Greek philosophers held a view which could be reasonably so called; but it remains a question whether this is true of other philosophical schools, and of people who were not philosophers at all. Of the theological scholars who have worked recently on time the only one who has published a thorough investigation of Greek theories of time is Delling, and I cannot find in him any statement that a cyclic view was common property to 'the Greeks'. Boman maintains that 'the popular time conception of the Greeks is as rectilinear

---

[1] Cullmann, p. 45 n.; E.T., p. 52 n. For some Greek views see, apart from general works on Greek philosophy, J. F. Callahan, *Four Views of Time in Ancient Philosophy* (Harvard, 1948), and works cited therein.

[2] Delling, p. 46.

[3] Boman, p. 123 f.; cf. p. 20, 143 f., 153.

[4] Of the particular phenomena cited by von Dobschütz, Delling (op. cit., p. 46) produces the quite opposite interpretation, namely that the Greeks felt so strongly their existence in time that they could not without great difficulty conceive a being apart from time; and that hence freedom from time for them had to lie in a world spatially far removed.

as our own', which might perhaps be true, even if one shudders at the argument from which it is supposed to follow (from the existence in the Greek verb system of temporal forms which can be marked off on a straight line).[1] Goldschmidt, a specialist working on the idea of time in Stoicism, mentions Cullmann's book with appreciation, but objects at once to precisely this point in it, that Cullmann accepts without discussion the idea that 'for Hellenism' the circle is the symbolic expression of time.[2]

This need not be further discussed here. Let it suffice that, if the lineaments of the biblical view of time must be seen by a contrast with those of the Greek view, it can by no means be taken for granted that 'the Greeks' in general entertained a cyclic view. Greater detail in investigation of Greek thought would be required, and further precision about what is implied in a 'cyclic view', and about the reasons why such a view is held at all, would also be needed.

Our second point is the question whether a 'cyclic view' of some kind cannot also be found in certain Hebrew sources. Here once again we have to turn to Qohelet. While it may be maintained that no excessive dependence can be placed on a book so untypical of the Old Testament canon, it remains true that this is the only book which can be said to have devoted an explicit discussion to time. It is extremely difficult therefore simply to rule it out, or with Wheeler Robinson to state roundly that its thought is 'un-Hebraic'.[3] The less the thought of Qohelet is

---

[1] Boman, p. 125, cf. p. 162.

[2] V. Goldschmidt, *Le Système stoicien et l'idée de temps* (Paris, 1953), p. 50 n.

[3] I should perhaps explain why I would be prepared to admit that Qohelet was 'untypical' of the biblical canon (an opinion which, surely, few would seriously dispute), and yet deny that he can be classed as 'un-Hebraic'. Surely this is not a serious contradiction. A person may entertain peculiar or anomalous opinions, and propagate them in writing, without being denied the title to belong to his own nationality and culture, and without his usage ceasing to be part of his own language system. An Englishman who spends his life in propaganda for the belief that the earth is flat or that the moon is made of cheese will no doubt be untypical of his nation, but is hardly to be classed as 'un-English'; and his English usage will still be quotable as part of the English language system, just as I have quoted the usage of Qohelet as evidence for Hebrew usage. Wheeler Robinson uses 'Hebraic' in fact in a schematic way, as if there was a homogeneous system of 'Hebrew thought', which could be used as a criterion by which a writer could be ruled 'un-Hebraic'. In fact when one gathers, as he rightly does, material which shows that Qohelet has no sense of history in the things he says, all one can reasonably do is enlarge our idea of 'the Hebraic' so as to include this Jewish thinker who wrote in the Hebrew language, and admit accordingly that there is no uniform system of Hebrew thought, about time as about other things.

held to depend on actual Hellenic influence (and such influence seems, in the judgement of recent research, to be rather unlikely), the more impressive is his insistence on the return of what has been before. 'Aller Nachdruck liegt auf der Wiederkehr.'[1] I do not suggest that Qohelet's view of time coincides with any Greek cyclic views of time; but it cannot be easily denied that it is in some sense a cyclic view, and that therefore the term 'cyclic view' cannot be reserved for products of Greek thought.

Thirdly, it may well be asked whether that which was cyclic for the Greeks (or some of them) was the same thing as that which in the Bible has been regarded as characteristically non-cyclic by Cullmann and others. Here Boman has made some valuable comments. When theologians say that the Hebrews did not have a cyclic view of time, they surely really mean that in Hebrew thought the sequence of historical events, or of some historical events, is a purposive movement towards a goal; it is certainly not cyclic in the sense of something recurrent, but is non-recurrent, non-reversible and unique.

But Greek 'cyclic' views are not necessarily opinions that historical events are recurrent.[2] What is called a 'cyclic view' on the Greek side is in many cases not a view that holds history to be cyclic, and in many cases has no particular connection with historical events. As Boman correctly points out, the emphasis on circular movement in Aristotle's doctrine of time arises from the perfection and uniformity of the circle, and has nothing to do with a circular disposition of the total succession of world events, that which according to Cullmann forms a straight line in Christian thought.[3] It is true that Aristotle refers to a current saying 'that human affairs form a circle', κύκλον εἶναι τὰ ἀνθρώπινα

[1] Galling, *Der Prediger Salomo* (Tübingen, 1940), p. 53. For other judgements about Qohelet's view of time see Sasse in *TWNT* i. 205; Vriezen, *An Outline of Old Testament Theology* (Oxford, 1958), p. 347. Knight, op. cit., pp. 87 and 337, like Wheeler Robinson, regards Qohelet as un-Hebraic and influenced by Greek philosophy, although the reference here is not to the subject of time in particular.

[2] It would appear rather that the idea of a recurrence of all historical events is a Stoic idea rather than general currency among Greek thinkers, although there were precedents, for example among the Pythagoreans. See J. von Arnim, *Stoicorum veterum fragmenta*, i. 109 (Zeno) and ii. 623-32, and M. Pohlenz, *Die Stoa*, p. 80 f. It is surely unlikely that the Greek historians expected an exact recurrence of the events they related.

[3] Boman, pp. 125, 162. In recognizing that Boman has made a good point here, I do not imply agreement with his arguments about time in general or with his final results about time. For Aristotle's passage see *Phys.* 223 b.

πράγματα, but this can hardly be called a 'cyclic view of time', for it is only a rather obvious judgement from the fact that human realities come to be and pass away, exactly the same point as occupied Qohelet. Apart from the uniformity of the circle, the other important influence in bringing forward the picture of time as something circular is surely the movement of the heavenly bodies by which time is commonly measured; so for example in the *Timaeus* the circularity of the movement of time, the moving image of eternity, is natural in view of the connection of the heavenly bodies with time (37 a-38 c).

But for this aspect of time in connection with the heavenly bodies there is no reason to doubt that the Hebrew understanding was cyclic exactly as that of 'the Greeks' was, and this not only in rather anomalous writers like Qohelet but in general opinion. Thus, for example, we do not have to resort to etymologizing interpretation in order to see that the sentences with phrases like *tᵉšubat haš-šana* 'the return of the year' (II Sam. 11.1; I Kings 20.22) and *tᵉqupat hay-yamim* or *haš-šana* 'the coming round of the year' mean in some sense a cyclic movement of seasons and natural periods. The writer of Isa. 29.1 asks 'let feasts come round' (*ḥaggim yinqopu*), and Job 1.5 tells how 'the days of the feast had completed their circuit' (*ḥiqqipu yᵉme ham-mište*; rendering of BDB). In so far then as Greek cyclic theories of time are based on the circular movements of the heavenly bodies, they are based on something which was quite naturally understood as cyclic by the Hebrews also.

Moreover, if we admit for the moment the validity, for discovering ideas about time, of the etymological methods which Orelli used and which have continued to be used in theological study, then we shall find ample evidence of cyclic views of time in Hebrew and in Semitic generally. Orelli himself did so, and devotes over ten pages to temporal terms derived from cyclic movement, including Hebrew *'od*, *tᵉqupa*, *dor* and *gil*, along with cognate words in other Semitic languages.[1] Those who are prepared to use etymology in order to show that *'et* 'time' means ' "occurrence", that which runs across us, meets us', and therefore 'time in the concrete, in its filled content',[2] must also be

---

[1] Orelli, pp. 30-41.  [2] Wheeler Robinson, op. cit., p. 109. His etymological survey on p. 120 f. does not include the words of 'cyclic' origin quoted by Orelli.

prepared to find from the same method, as Orelli did, ample evidence of cyclic time conceptions in Israel. Many writers seem indeed to have ignored this aspect of Orelli's survey, in spite of wide acceptance of other points made by him.

To summarize this point, then, we may say that the discovery of a cyclic view of time among Greek thinkers does not necessarily in itself constitute an opposition to Hebrew thought. A definite contrast would appear if a cyclic view of *history* were found, so that all historical events would be expected to recur. A valid contrast may also perhaps be made by pointing to the absence of any great significance for *history* in Greek metaphysics. But it cannot be assumed without much greater care and without detailed justification that the appearance of something that could perhaps be called a cyclic view of time on the Greek side, or of something that could perhaps be called a linear view on the Hebrew side, in itself constitutes a fundamental contrast between them. In general it might be wise to abandon the phrase 'cyclic view of time' and in all cases specify such different possibilities as the following: (a) the circular movement of earthly existence, coming to be and passing away, (b) the circular movement of the heavenly bodies which measure or govern time, (c) the application to time of circularity as the example of uniform motion, (d) the idea of a cycle of cosmic process occurring only once but ending up where it began, (e) the idea of a cycle of cosmic process repeating itself infinitely, (f) the idea of a temporal cycle in which all historical events recur as before. These are certainly not all the same thing.

Assuming also that we must continue to have the generalizations about Hebrew and Greek thought which have lately been so common, much greater care is needed in formulating them. Quite apart from the dubious lexical methods which have been used, the generalizations themselves vary very widely from one author to another; so much so that one is repeatedly asking on what basis they are made and whether some method might not be stated by which one might have some hope of determining which of them may be true.

Thus Cullmann, to begin from the best known, regards the biblical way of thought about time as 'linear', and thinks this to be its great difference from the Greek. Delling, on the other hand, who in certain respects is quoted by Cullmann with apparent approval,[1] says it might be possible to say of Saint Luke that his understanding of time was essentially linear and thus Greek.[2] Von Rad maintains that 'the conception of time in which the Western man more or less naïvely lives is linear', in this respect apparently presenting a strong contrast with the biblical conception.[3]

To Marsh, on the other hand, while the cyclic view of time is 'Greek', the linear is not biblical either, being produced only by the false importation of (Greek?) 'chronological time' into the understanding of the biblical material. To him there is really no biblical view of time but only one of decisive times or moments[4] —a position to which Cullmann himself at one stage comes very near, and which he avoids only by his very dubious argument that I Tim. 2.6 represents the process of forming a line to join up the 'kairoi'.[5] Boman, again, seems to maintain that the Greek view of time could include both the linear and the cyclic, but that both of these are a spatialization of time; while for the Hebrews time had nothing to do with space and was never represented through the use of spatial images—a position which depends on quite fantastic explanations of linguistic phenomena in Hebrew.[6] Buess says that neither the linear nor the cyclic aspect is characteristic for the New Testament idea of time; it is rather 'something like a *cross of time (ein Zeitkreuz)*, which subjects to its own rule the cyclic motifs also'.[7]

Pidoux on the other hand hospitably admits that 'there is no single biblical conception of time, but the Bible contains several'.[8] Eichrodt, conversely, is willing to admit that there may be none at all, or none in the sense of a view of time which is quite distinctive to the Bible. In what is perhaps the most moderate and sensible treatment the subject has had in recent theological work, he suggests that there may be no biblical conception of

[1] Cullmann, p. 44 n.; E.T., p. 51 n.    [2] Delling, p. 81.    [3] Von Rad, p. 112.
[4] Marsh, *Fulness*, pp. 175-9.    [5] Cullmann, p. 36; E.T., p. 43.
[6] Boman, pp. 123-83. See my criticisms of his linguistic arguments in *Semantics*, pp. 46-88.
[7] E. Buess, *Die Geschichte des mythischen Erkennens* (Munich, 1953), p. 152. On p. 151 Buess renders a useful service by showing that the view that mythology always depicts time as cyclic is a product of the 'Pan-Babylonian' school, and one containing many weaknesses.    [8] *RThPh* ii (1952) 125.

time which involves a position towards time that is really opposite to our own and that arises from a quite different understanding of reality. He holds that the important thing for the Bible lies not in the idea of time itself but elsewhere, in the use made of the historical sequence for the presentation of an encounter with God.[1]

This extreme variety among a series of authors, all of whom write within the context of modern biblical theology and accept its general methods, and all of whom are interested in noticing and bringing out elements which are unique to the Bible in its conception of time, suggests that there is some severe fault in the methods used, or that there simply is not material from which such questions can be decided, or that the writers have not been able to avoid a very considerable confusion in expressing themselves and in understanding one another.

As a corollary to the above section one thing may be pointed out. The difficulties and uncertainties have in large measure arisen from the assumption that a contrast with *Greek* thought will be a proper way in which the true character of the biblical thought may be brought out. This assumption, which is extremely common, and indeed more or less standard, in modern biblical theology, itself involves highly questionable assumptions about cultural comparability; but this will not be further discussed here. Only one particular difficulty will be noted here: some of the problems and contradictions arise from the attempt to compare *formulated* Greek opinions about time, stated and discussed at length by Plato or Aristotle, with Hebrew views which can at the most be called implicit and unformulated.

We now turn to consider one or two ways in which, even if we do not believe that the Bible furnishes a view or doctrine of time, we may find that a starting-point or basis of some kind is afforded in the Bible.

One of the more important general problems in the theological-philosophical area is the relation between time and eternity. Here, as we have seen, Cullmann maintains that eternity is not something other than time, and in particular that it is not a

[1] *ThZ* xii (1956) 125.

timelessness of any kind. Marsh's doctrine on the other hand is that the eternal, although by no means identical with temporal succession, is not to be found through the negation of time, and indeed contains within itself as an essential element its own ingression into time which takes place in the Incarnation.[1] We have seen that Cullmann's argument from the terminology, which, put at its simplest, is that, since the same word is used in the Greek New Testament both for eternity and for certain periods of time, therefore eternity is not qualitatively different from time, is a quite invalid one. This however does not in itself prove that Cullmann is wrong in holding eternity to be unlimited time, and not something other than time, for it may be that there are philosophical and theological reasons for holding this even if the linguistic ones are faulty. We may therefore suggest one or two points which may be worthy of consideration in the more general philosophical-theological area, but still assuming a desire to relate what is said in this area to biblical data.

Firstly we may recall a point which has been briefly mentioned,[2] namely the importance of the biblical creation stories, and especially that of Genesis 1. It is essential to Cullmann's case to deny that time began with the creation of the world. But we have seen that, though it is hardly expressly asserted that it did so begin, there is a fair or considerable probability that the passage meant, or was understood to mean, something not wholly contrary to such an assertion. It would perhaps be possible in theory to maintain that before the creation of the world there was not something other than time, but time of another kind. But it would be difficult or impossible to produce biblical evidence for such an idea, and it would be irreconcilable with Cullmann's demand that we must take the most naïve and unphilosophical approach possible—a demand which may be reasonable for us when we try to put ourselves in the place of the writers or the early readers of Genesis, whether or not it is a good principle for the modern theologian trying to work out the truth about time. If, then, there is some reason to suppose that Genesis meant that time began with creation, then there is at least some case for talking about 'eternity' as a reality other than time. The following up of this theme could hardly find a decision from within the

[1] Marsh, *Fulness*, p. 139 ff.    [2] See above, p. 75.

Bible; but it would have *started* from a point in the Bible; and this would be not a lexical point, but the probable meaning of a passage. It is possible therefore (I do not assert more) that this might be a biblical starting-point for a discussion of eternity as something other than time.

Secondly, we have seen that Cullmann's argument is faulty because it works from a lexical structure of hypostatized words. We have also seen that the Bible appears to offer no explicit group of statements expressly devoted to the problem of the relation of time and eternity—the fact which drove Cullmann into his artificial lexical structure in the first place. In the actual contexts the phrases including the word αἰών are used in relation to other realities, such as God, his works and words, and the institutions set up by him. It may be helpful to concentrate not on the word αἰών but on those realities with which the phrases using it are connected.

(a) In some cases only a remote point in time within human history, and the interval between that point and the present, is involved, as in the οἱ ἀπ᾽ αἰῶνος προφῆται of Luke 1.70. In such a case, even if 'eternity' were used as a translation, which would hardly be likely, the reality involved would not qualify for a discussion between 'the totality of time' and 'something other than time', since neither of these alternatives is possible here.

(b) Secondly, there are cases where the result of some divine act, or some institution set up by God, is expected to last 'for ever' or 'unto eternity'. In these cases again the alternatives of 'the totality of time' and 'something other than time' do not present themselves for choice, since the act or institution as described is stated to have had a beginning in time.

(c) Thirdly, there are passages mentioning God himself, the glory due to him, the unseen world, the everlasting fire, and so on, where it is extremely likely, if not certain, that the realities mentioned are taken by the writers to have been existent as such throughout the totality of time. Now while perhaps (or certainly) the writers would understand these realities not to be in a static condition throughout the totality of time, but to be living and active, it remains true that continuance in the same status, of being God or of being fire or whatever it is, throughout the totality of time, is something rather different from the condition of other beings which are ephemeral. In other words, when the writers say that God is 'eternal', they may well be saying rather more, or meaning rather more, than that his existence is conterminous with the totality of time. We may therefore consider it meaningful to speak of 'eternity' as a useful designation for the kind of existence involved in these cases. And if this is granted, it is possible that the same usage should be extended to those institutions set up by God which have an origin in time but which thereafter are expected to last 'for ever' or 'unto eternity'.

It might indeed be pointed out that such a group of realities, in relation to the existence of which one might meaningfully speak of 'eternity', would include many of those which in the New Testament are qualified by the adjective αἰώνιος. But this of course affords no precise demarcation of the group. When we form such a group we may as well be aware that we are leaving the area where a demarcation of realities can be strictly correlated with the appearance or non-appearance of any one biblical word. Such a group would be demarcated, not by the exclusive and peculiar use for it of any one designation, but by the fact that other aspects of biblical thought lead us to suppose that these entities form a special class. There is in fact no reason why we should not form conceptual demarcations which do not correspond with demarcations between units of the biblical lexical stock, just as (as we have seen from the case of καιρός and χρόνος) we can go far wrong by trying to form our conceptual demarcations by basing them upon lexical demarcations. Thus surely no one will suppose that because the same word αἰώνιος is used of the Jerusalem temple gates in πύλαι αἰώνιοι (Ps. 23(24).7) and of the Christian God in Rom. 16.26, therefore nothing different is intended in the relation of the gates to time and of God to time? Yet arguments no better than this have been current coin in the lexical method of Cullmann and others.

It must be clear that no kind of theology can be built upon such material as the occurrence of this word in each of these places, or the occurrence of different words in two places. A valid biblical theology can be built only upon the *statements* of the Bible, and not on the *words* of the Bible. On the other hand, there may be distinctions which can be validly observed within the thought and expression of the Bible, but which do not coincide with differences marked by the use of different lexical items; and there may be good reason to follow up these distinctions. It is on this level, most probably, that some discussion of the relation of time and eternity on a biblical basis might be usefully re-opened.

When we say, however, that theology must be built upon the *statements* of the Bible, we raise the question whether Cullmann in particular has been asking a meaningful question at all. In his preface he states that his interest will be entirely directed upon that which is '*specifically* Christian in the New Testament revelation, in other words precisely that which it does not have in common with philosophic or religious systems'.[1] Such material is to be found in connection with time, and we hear of 'really basic assertions of New Testament theology about time'.[2] But where are these basic assertions? If, as Cullmann appears to maintain, one of them was that time was a straight line, and another that eternity was not a condition of timelessness but the unlimited totality of time, why is it that in the New Testament records of the preaching to the Gentiles there is not a trace of any such assertion having been made? St Paul at any rate had the mental and linguistic equipment to make them, and in a number of passages points out difference of thought between the Greeks and the Jews, or between the Greeks and the Christians; yet we have not the slightest evidence that he ever said:

'You think time is a circle, but we think it is a straight line' or
'You think that eternity is timelessness, but we think it is the totality of time'.

The conclusion that the conception of time forms a basic assertion of New Testament Christianity seems therefore very ill-founded. It may well be suggested, perhaps, that a different conception of time was *presupposed* by the early Christians from that which was customary among their Hellenistic neighbours. But even so the fact that, in the records which we have, this supposed presupposition was never brought to light, expressed and explained to Hellenistic audiences and Hellenistic converts

[1] Cullmann, p. 8; E.T., p. 12.    [2] Ibid., p. 15; E.T., p. 19.

makes it difficult to believe that its importance was very great.

There remain one or two other possibilities for a biblical approach to time. One is of course to work from Qohelet, which is, as J. A. T. Robinson correctly states, 'the only book in the Bible consciously exercised by the *problem* of time'.[1] But it is doubtful if even our strongest enthusiasts for 'biblical thought' are really prepared to base a doctrine of time on what this writer says—a fact which itself constitutes a serious challenge to the prevailing conception of 'the unity of the Bible'.

Further, it may be possible to state something by a literary study of biblical literature in its approach to historical narrative or to general human questions. This study would not work from formulated views of time (for they do not exist), nor from lexical material, but from stylistic criteria and general literary methods. Such a study would use the same methods and have the same validity and range as one of, let us say, 'Homer's view of time' or 'Herodotus's view of time'. It is possible that interesting and valuable material, within its limits, could thus be assembled. But its significance for theology would be limited in the following ways. Firstly, it would be likely that substantially different results would be reached for different biblical writers, and improbable that a solid 'biblical view' could be obtained. Secondly, such aspects of stylistic and literary form could not be disengaged from the literature in which they are found, and 'used' by being integrated into a doctrinal statement of what time in fact is. They would be discovered expressly by *our* asking certain questions of the style and expression of the writers in their handling of certain matters (other than time itself directly), and they would be valid only within the context of the writings where they were observed. Such stylistic points might be imitated (e.g. in hymns or in creative writing), but could not be formulated or used in doctrinal argument or formulation.

The position here developed means in effect that if such a thing as a Christian doctrine of time has to be developed, the work of discussing and developing it must belong not to biblical but to philosophical theology.

In fact we may observe that the origin of the question lies in

[1] In *Theology* lvi (1953) 107.

many ways within philosophical rather than biblical material. We cannot take seriously Cullmann's insistence that all philosophical considerations must be excluded, for the question he asks, namely 'What is the nature of time and eternity in biblical thought?', is a question for which the Bible itself gives no precedent, and one for the answering of which it affords so little material that his appeal has to be one to the lexical stock of the Bible rather than to its actual statements. Moreover, for Cullmann himself we can probably say that the urge to investigate the whole matter has arisen not simply from a descriptive study of biblical material but from a sense that certain general theological-philosophical problems can be solved, and the disagreements among exegetes in certain respects overcome, through a special concentration on the subject of time.[1] Something similar can be said of Marsh. It is clear throughout his book that one of his main interests is to find a formulation of the centrality of history for Jewish-Christian faith, while recognizing the fragmentary and inadequate nature of the biblical records on the one hand and avoiding a crassly chiliastic understanding of historical process on the other. The same is true of Robinson. In the case of Orelli the use of a kind of philosophical starting-point and organization was explicit, and has been brought out in my discussion of him. Ratschow's article includes a good deal of non-biblical philosophical material, and this makes difficult the application of his discussion to the biblical material and also shows the contradictoriness of combining them. If, then, many or most of the modern discussions start out in fact from some problem or difficulty found in modern theological-philosophical discussion, we may regard it as probable that satisfactory results can be reached only by seeking clarification within that area, and not by expecting the Bible to answer the problem for us. If, on the other hand, Eichrodt is right in suggesting that there is in fact no biblical notion of time which differs widely from our own, this may be a reason for theology to avoid being forced into developing such a theological doctrine of time, or at any rate to avoid claiming that any such doctrine developed rests on a certain biblical basis.

[1] See for example Frisque, pp. 240 and 57-63, who well brings out the particular importance for Cullmann of the contrast with Bultmann.

Not only has the modern question about time in theology taken its rise in many cases from philosophical-theological questions; it can also be said that many of the solutions show dependence on modern theological-philosophical formulations, adapted or re-moulded for the categorization of biblical material. Thus it will surely be thought likely that the distinction of two kinds of time, 'realistic time' and 'chronological time', which forms the basis of the treatments by Marsh and Robinson, is at least partly suggested and occasioned by the distinctions set up by Bergson for the same subject. This is even more true of Tresmontant, of whom Frisque[1] rightly says that he identifies the biblical emphasis upon history too easily with Bergsonian philosophy. Boman too depends very considerably on Bergson in his identification of what he supposes to be Hebrew thought, although one must not attribute to Bergson himself the imaginative re-shaping of his philosophical concepts by Boman in order to apply them to Hebrew thought.[2]

On the other hand it may be suggested that Cullmann's main position, namely that eternity is not other than time, but is time without limit, has clear Aristotelian affinities. It is the result which Orelli reached when he started out from an explicitly Aristotelian philosophical basis for his philological research, and it is noticeable that Cullmann's polemic against 'Greek ideas' or 'philosophy' is very commonly in fact against the *Platonic* conception of a timeless eternity.[3] In this connection we may mention also that in Cullmann's thought time itself comes to be, as in Aristotle, something without beginning or ending. This affords us the delightful spectacle of Karl Barth being charged with Platonism for holding the contrary.[4]

We may conclude therefore that the prominence of philo-sophical-theological considerations, both in the posing of the problems about time and in the solutions offered, suggests that the question should be handled much more frankly and more completely within the area of philosophical theology,[5] and that it

---

[1] Op. cit., p. 237 n.    [2] Boman, op. cit., pp. 22, 126 f., 129, 142, 144, 174, 194 f.

[3] E.g. Cullmann, pp. 52-59; E.T., pp. 61-68. The similarity of Cullmann's view to Aristotelianism is noticed also, and independently of the present writer, by J. McIntyre, *The Christian Doctrine of History* (Edinburgh, 1957), p. 23 n.

[4] Cullmann, pp. 54 f.; E.T., p. 62 f.

[5] One may indicate as an example of possible lines of progress the treatment of history by McIntyre, already cited.

is largely failure to be sufficiently self-critical within this area itself that has led to the attempt to find within the Bible the correct posing and answering of the questions.

# VII

## CONCLUSIONS FOR BIBLICAL INTERPRETATION IN GENERAL

IT remains to suggest certain conclusions which may arise from our study and which may be of value in our approach to biblical interpretation in general. As stated at the outset, this study was embarked on, not with the intention of establishing the biblical view of time, but in order to establish through the examination of a well-known subject certain important positions for method in biblical study. It is true that our study may among other things provide some new material and some new principles of organization for those who wish to pursue farther the establishment of a doctrine of time on a biblical basis; on the other hand it has been suggested, and not from some extra-biblical position but on the basis of the biblical material itself, that there may not in fact be sufficient material in the Bible on which a purely biblical view of time may be built. Our main purpose has lain elsewhere, however: it is to examine and criticize certain methods which are not uncommon, in order to gain some lessons for the better handling of biblical evidence. For this the subject of time seemed excellently suited, because of its general interest and of the number of works which have tried in recent years to approach it on the basis of biblical evidence.

First of all, then, it has been shown that the use of lexical structures in the investigation of time has been extremely faulty. While this faultiness may in part be traced to the overlooking of certain evidence, I submit that the method itself is faulty, and that the overlooking of evidence necessarily followed from the adoption of the method, because if the evidence had been carefully assessed it would have only demonstrated the inviability of the method. Certainly the fact that the method is faulty in the case of time does not show that it must always be faulty, and there may be cases where the circumstances are different and where the following of some such method might be much more justified. Such cases might appear where a pair of words, or more pairs than one, form within a certain area a recognized and technical system by which relevant realities may be classified.

One thinks for example of the two pairs 'holy' and 'profane', 'clean' and 'unclean', within the priestly system of holiness in the Old Testament. But one is not entitled to treat in this way any pair or group of words relating to some reality, without special justification. In the cases of καιρός and χρόνος, and of καιρός and αἰών, we have seen that the method was in principle unjustified and in practice faulty.

Even where the treatment of groups of words as a kind of structure may have some justification, it has to be guarded against certain dangers which have been disclosed in our study of treatments of time. One such danger is the domination by etymology of the organization and conduct of the study, a situation which is most marked in Orelli. Although many modern lexical treatments in theology have succumbed to this danger, it has fortunately been avoided in the treatments of time.[1] A more serious danger is the loss of control of the syntactical environments in which the words occur; the loss of perception, for example, that the question whether καιρός contrasts with χρόνος in New Testament Greek depends entirely on the syntactical environment. Inasmuch as 'the meaning of a vocabulary item is a function both of the item itself and of the item as occurrent in various contexts',[2] all lexical treatments which neglect the determinative importance of the syntactical context risk misrepresentation of the facts. This risk is present with all arguments of the type which say 'the word here is καιρός, and therefore a critical time is intended', and similarly for other words.

Modern biblical theology has rightly made it its concern that biblical terminology should be seen 'within the context of' biblical thought as a whole. But it is a misuse of this principle to apply it to words, in such a way that the relating of the word to features of the general context of biblical thought replaces the examination of its actual syntactical context. The general theological context can never in the slightest degree be a substitute for the syntactical environment. Only within their syntactical environment do words function. Where this is neglected, we may produce studies which quote Greek and Hebrew words in

---

[1] We must be glad that more has not been made by theologians of such suggestions as that καιρός may come from the same root as κρίσις, *cerno, discrimen*, etc.; Pfister, 'Kairos und Symmetrie', p. 131 ff.

[2] From R. E. Longacre, in *Language* xxxii (1956) 302; quoted in *Semantics*, p. 38.

every sentence and which clamorously insist on the pursuit of biblical terminology, but which in fact are not dealing with biblical language at all.

All this must be borne in mind as a necessary compensation for the popularity of word-books and theological dictionaries as instruments of biblical interpretation at the present day. Such works run the risk of the faults just described, or, where they themselves avoid these dangers, they may nevertheless allow their readers to fall into them. In general, their very presence and nature tends to suggest that, as Cullmann in his own study put it, 'the New Testament terminology for time teaches us that' such and such is the case. To this one must say that it is dubious whether biblical terminology ever 'teaches us' anything in this sense. Linguistically, it is the syntactical complexes, in which the lexical items are used, and not the lexical items themselves, which constitute communication. Theologically, it is the communications made in the Bible, and not the lexical stock used in them, that 'teach us' the truth about God or about sin or about redemption. Fortunately, we do not need to devise any very revolutionary procedures to correct these faults where they appear. One procedure, the attention to translation, has already been mentioned. Another, the writing and use of commentaries on the text, is after all the traditional form of biblical interpretation. The present tendency for word-studies to become a kind of rival to commentaries for the centre of interest in biblical interpretation is an unfortunate one. When a theological word-book states that it 'owes its conception to the conviction that the words of the Bible are not merely interesting objects of academic research, but are indeed the words of eternal life', we must simply and firmly state that the words of the Bible are nothing of the sort. The 'words of eternal life' are not the lexical units of the biblical languages, but the sentences and speeches and narratives of the Bible. The sentence quoted depends on a tacit passing from one sense to another of the word 'word'.[1]

In this connection one particular point may be noticed. The

---

[1] Those who, after all my strictures, are still willing to derive 'teaching' from the lexical layout of Hebrew will not fail to notice that two quite distinct and unrelated words would be used in Hebrew for the two senses of 'word' in the sentence quoted, since the kind of 'word' which one finds in a word book is *milla*, while the kind which appears in 'words of eternal life' would probably be *dabar*.

present prominence of lexical studies is probably in part due to the ease with which they can be integrated with the teaching of dogmatics. This is especially so where the layout of words in a lexical stock is supposed to form a pattern identical with the real structure of the reality, God or sin or time, which dogmatics have to handle. In this situation faulty conceptualizations and hypostatizations of lexical items can easily occur. But apart from this, it would not be doubted that in many cases dogmaticians simply do not know the biblical languages, and especially Hebrew, well enough to be able to assess the syntactical factors in the meaning of words. To have in one's mind a list of the Hebrew words for the main theological realities is not evidence of ability to handle a Hebrew sentence in which some of them occur. This fact on the one hand makes more attractive lexical procedures in which the more purely lexical aspects of biblical language are supposed to 'teach' theological truth, and in which syntactical and other factors are neglected. On the other hand it naturally leads to the type of error which arises from neglect of the syntactical environment. The corrective for this difficulty is a greater reserve towards such lexical methods, and on the other hand a greater use of English or German words, and not of Hebrew or Greek words, in the more strictly dogmatic discussion of the realities handled. This again is no revolutionary suggestion; we have already seen that substantial dogmatic treatments of the subject of time have already been offered, for example by Barth and Brunner, which make very little reference to the Hebrew and Greek words of the biblical tradition.

Such then are some limitations and inadequacies of the lexical methods which have become fairly popular in theology. In the introduction it was suggested that the general approach to biblical interpretation, popular in modern theology, would not be criticized except in so far as it depended upon the use of this sort of lexical method. How far in fact has this turned out to be the case, and how far therefore does our study suggest that prevalent methods of handling the Bible require modification?

Firstly, if a biblical view of some subject can be extracted from the Bible only by the use of these lexical methods, it may well mean that we have to abandon the hope of handling such subjects on a purely biblical basis at all. If we follow Cullmann in his strict adherence to a biblical basis and his rigorous exclusion of philosophy and of traditional theology beyond the biblical area, and if this forces us into the lexical methods here criticized as the only means of reaching biblical thought, then the alternative is to recognize that the subject is not a genuine biblical theme, and that on the principle of adherence to a strict biblical basis nothing can be said about it. In other words, our criticisms lead here not to an opposition to the idea of biblical theology, but to a limitation of its scope. This is surely no loss to it, for we are not limiting its scope artificially, but limiting it to the subjects actually handled in the Bible. After all, as we have seen, the apostles never enunciated in their preaching, as far as we know, a doctrine of time. Even if some different subconscious approach to time was behind their thinking, how can we suppose the statement of this approach to be an essential for the communication of the Gospel, when they in fact communicated the Gospel to men of all nations without in fact making this statement?

This limitation is of vital importance to biblical interpretation, for by means of it we can remove from the forefront of interest assertions which obscure the centrality of other assertions. If we are to look for what is specifically Christian in the New Testament, to find the really basic assertions of New Testament theology—the purpose which Cullmann announces at the beginning of his quest—surely we should turn to such great themes as the passion and resurrection of Jesus Christ, and the forgiveness of sins through him—all themes to which Cullmann himself gives full value. But surely it is intolerable to place in the same category, and class also as specifically Christian and fundamental, such other assertions as the assertion that time is a line, and that eternity differs from it only by being unlimited! It is not only linguistic method, but also theological acceptance of biblical authority, which requires that biblical interpretation should concentrate primarily on the communications actually made by the men of the Bible, and not on what can supposedly be read between the lines of the Bible.

But this has some effect on the employment of the categories of the unity and distinctiveness of the Bible, which, as we have seen, would be widely agreed to be fundamental to much of the present theological approach to the Bible. Let us look first at its unity. That the Bible is a unity in the sense that it forms a written crystallization of one great tradition, and that for the men of the New Testament the tradition of the Old Testament was basic and normative, is surely incontrovertible, and seems to the present writer fundamental. But this unity has been mis-stated or misused as follows. Firstly, since within this unity of tradition very considerable differences in actual teaching, doctrine and practice occur, there has been an increasing tendency to state and evidence the unity of the Bible by pointing to those features of the biblical mind which are not the subject of actual biblical assertions, but which may be held at the most to be implied by the Bible, or may perhaps be deduced from lexical studies of its terminology. Secondly, excessively strained efforts have been made to connect this unity of biblical tradition with the presence of one roughly unitary language system for the Old Testament and another for the New; and, moreover, to make these two as far as possible into one by emphasizing the dependence of New Testament Greek upon Hebrew. This second aspect has greatly stimulated the use of lexical procedures, because it is only to a very limited extent in morphology or in syntax, and therefore almost entirely in lexical semantics, that an influence of Hebrew upon New Testament Greek can be traced. Our criticism of lexical procedures in this book may suggest that these ways of stating or using the unity of the Bible require to be modified.

Similar considerations apply to the distinctiveness of the Bible. Firstly, the distinctiveness of the Bible should not be so stated as to suggest that, where some point of view about a sub-ject exists outside the Bible and in an environment, such as that of Greek philosophy, which strongly contrasts with the Bible, the Bible will certainly contain or imply a view of that same subject entirely distinct from that found outside of the Bible. The fact that the Greek philosophers had theories about time does not mean that the Bible will necessarily have a view of time which contrasts strongly with theirs, or indeed have a 'view' of time at all.

Secondly, the distinctiveness of the Bible may be stated in a quite distorted way by insisting that it is against *Greek* thought that its distinctiveness will be properly displayed.[1] (a) It is only to a very limited extent that the conflicts of ancient Israelite religion, which dominate the Old Testament, were against Hellenism. (b) One can often make an easy case for the distinctiveness of the Old Testament by contrasting it with Greek thought, but this only leaves open the possibility that it will not contrast similarly with non-biblical Semitic thought, such as that of the Phoenicians or the Arabs. (c) It is true that the contrast with the traditional thinking of the Greek world is important for the understanding of the New Testament, most of which reached its present form in that world. But here we should surely be guided by the New Testament itself. In contrast with the present theological scene, the New Testament is notable for the limited and partial character of the criticisms which it made of Greek thought, in contrast with the comprehensive and systematic contrasts offered by modern theology. The apostles do not seem to have seen that they could materially assist their hearers or strengthen their case by pointing out that Greek thought is static and biblical thought dynamic, or that biblical thought is concrete while the traditional thought of their hearers has been abstract, or that the many misunderstandings and schisms in the Church could be largely attributed to the presence of such traditional Hellenic distinctions as those between form and matter, or appearance and reality. The systematic and comprehensive series of contrasts, with which much modern theology sets biblical thought against Greek thought, is something quite unlike the actual presentation of the Christian message in Hellenistic lands in New Testament times.

It may be thought that to admit this is to open the gates to the 'Hellenizing' type of New Testament interpretation, which regarded Paul and others as drawing much of their material from the mystery religions and the like. Nothing is farther from the intention of the writer, and to him the theory of a substantial influence of such forces on men like Paul is a singularly improbable one. The influence of this kind of Hellenistic life on Paul is not

---

[1] It is thus surely a curious contradiction, yet not untypical, that Cullmann, while strongly insisting on the complete contrast of the biblical view of time with the Greek, and indeed with all religious and philosophic systems, finds it natural to admit that Iranian religion agrees with biblical in seeing time as a line; p. 44; E.T., p. 51.

to be compared with that of the Old Testament, of his own Jewish background, and of the life and ministry of Jesus Christ. It remains true that it is a distortion to suggest that an *opposition* to Greek thought is a key to New Testament interpretation, and when such an opposition is singled out as the basic means of displaying the distinctiveness of the Bible it means the adoption of a procedure which has no direct justification from either the Old Testament or the New.

Thirdly, like the unity of the Bible, its distinctiveness can be wrongly stated and badly used when it is applied to the language systems of the Bible. These systems are not distinctive in the sense in which biblical religion can be claimed to be distinctive, and are not distinctive except in the sense in which all language systems are distinctive. No matter how strongly we accept the position that the Bible is theologically distinctive, we do wrong to try to correlate this with a corresponding distinctiveness in biblical Hebrew or in biblical Greek. It is little more than a superstition to suppose that Hebrew is as completely unique and unparalleled a language as Christianity is a unique faith. It requires very strained argument to maintain for a large area of New Testament Greek that its words have meanings divergent from those of non-biblical Greek in the same measure as that in which biblical thought differs from Greek philosophy or the like; that such divergences do occur in a fair number of places is no sign of *special* distinctiveness. Finally, and most important for our subject, the lexical stock of neither Hebrew nor New Testament Greek is laid out in a plan or pattern which corresponds with the distinctiveness of biblical thought.

In these respects, then, our study of the recent treatments of time suggests that ways of presenting the distinctiveness of the Bible require modification.

These considerations about the statement of the unity and distinctiveness of the Bible mean a very considerable modification in the positions which have been characteristic of modern biblical theology. The question which arises from them, and which will not be pursued farther here, is whether anything recognizably like our recent biblical theology can survive once these modifications are made, in other words, whether the making of these necessary modifications is not likely to be fatal to anything reasonably like the position which now stands as that of biblical theology. This question may well, however, be left to be answered after longer experience.

One reminder should however be given here. In so far as certain theological positions (and that of modern biblical theology is certainly a case in point) claim to stand upon, and to do best justice to, modern biblical and linguistic scholarship, the discovery that these positions depend upon, or tend to produce, repeatedly and systematically faulty assessments of linguistic evidence is finally and incontrovertibly fatal to these positions. Against this fact, if it is a fact, the general theological attractiveness of the positions, or the fact that they have done much good by correcting other wrong emphases, or the fact that the Church has found them rather helpful, count for nothing. Good scholarship and good theology alike require that where an interpretative method is found to be systematically faulty no pains should be spared to make the necessary adjustments.

The basic fault in the material we have been criticizing is that assumption which the writer has already criticized in his *Semantics*—the assumption that biblical language, in its grammatical mechanisms or in its lexical stock, will somehow surely and naturally reflect, or correspond to, or cohere with, biblical thought. For the criticism of this assumption, apart from the general considerations which I have presented elsewhere, two points may be made here. Firstly, a meticulous observation of actual usage in all its details is necessary in the student, since only such a meticulousness is sufficient to overcome the attraction of those aspects of fact which can appear to 'fit' with the structure of biblical thought and theology. Secondly, for the analysis and correction of the bad habits in linguistic interpretation which have lately become so common, methods may be required which would not necessarily be normally required in biblical interpretation. The considerable armoury of linguistic material which has been assembled in this book is not intended to impress people, or to suggest that biblical interpretation cannot be undertaken without Syriac and Coptic and the like. On the contrary, since biblical interpretation in theology must work from the things said in the Bible, and not from the lexical resources used in saying them, the fundamental points of biblical assertion will normally be visible to those who do not know the original languages. Those who do know them will be able to understand the text with very much greater accuracy, provided that their knowledge of the original language is in fact sound, careful and accurate in detail; but it is unlikely that in more than a few special cases this knowledge will lead to a recognition of some biblical conception which is vital to the understanding of the Bible, but which is quite invisible to the reader of the English Bible, because it is tied to the layout of the Greek or Hebrew lexical stock. If such cases were both numerous and important, one need hardly remark, it would be a poor prospect for the average layman—and indeed, one may add, for the average minister, whose Greek and Hebrew are not always excessively fluent. Theological assertions, which are based not upon what the Bible says but upon the linguistic resources which it uses, must be regarded with caution; their theological authority must be doubtful, and their linguistic validity must be carefully tested by satisfactory linguistic criteria. The use of such criteria may involve types of linguistic argument and description which have not hitherto been familiar in the theological area, and some of these have been rather roughly applied in this study.

But the main principles of this present study are not new. They are: firstly, the final authority of usage in all matters of linguistic interpretation, and the impossibility therefore of over-riding it with considerations drawn from the patterns of biblical thought; secondly, the essential place of the syntactic environment in the functioning of lexical units—something into which a real insight is given by our knowledge of the original languages, and which therefore provides a main justification for the work expended in gaining that knowledge; and thirdly, the principle that theological interpretation of the Bible must be not interpretation of its linguistic structure or of the layout of its lexical stock, but interpretation of the things which it says or asserts: linguistically speaking, of its sentences and larger complexes of speech. The observation of these principles has always been the foundation of sound exegetical work.

# ABBREVIATIONS

| | |
|---|---|
| *ABR* | *Australian Biblical Review.* |
| *BAss* | *Beiträge zur Assyriologie.* |
| BDB | F. Brown, S. R. Driver and C. A. Briggs, *Hebrew and English Lexicon of the Old Testament* (Oxford, 1907). |
| *EvQ* | *Evangelical Quarterly.* |
| *GGA* | *Göttingische gelehrte Anzeigen.* |
| GK | *Gesenius' Hebrew Grammar* (ed. E. Kautzsch, 2nd English ed. by A. E. Cowley, Oxford, 1910). |
| *HBAT* | *Handbuch zum alten Testament.* |
| *JBL* | *Journal of Biblical Literature.* |
| *JBR* | *Journal of Bible and Religion.* |
| *JEA* | *Journal of Egyptian Archaeology.* |
| *JJS* | *Journal of Jewish Studies.* |
| *JSS* | *Journal of Semitic Studies.* |
| *JTS* | *Journal of Theological Studies.* |
| KB | L. Koehler and W. Baumgartner, *Lexicon in veteris testamenti libros* (Leiden, 1953). |
| KD | Karl Barth, *Kirchliche Dogmatik* (Zollikon, from 1932); E.T., *Church Dogmatics* (Edinburgh, from 1936). |
| LS | H. G. Liddell and R. Scott, *Greek-English Lexicon* (revised ed. by H. S. Jones and R. McKenzie, Oxford, 1940). |
| MM | J. H. Moulton and G. Milligan, *The Vocabulary of the Greek Testament* (London, 1914-29). |
| *PL* | *Patrologia Latina* (ed. Migne). |
| *RhM* | *Rheinisches Museum für Philologie.* |
| *RThPh* | *Revue de théologie et de philosophie.* |
| *SJT* | *Scottish Journal of Theology.* |
| *StTheol* | *Studia Theologica.* |
| *ThLZ* | *Theologische Literaturzeitung.* |
| *ThZ* | *Theologische Zeitschrift.* |
| *TWNT* | G. Kittel (ed.), *Theologisches Wörterbuch zum neuen Testament* (Stuttgart, from 1933). |
| UPZ | U. Wilcken, *Urkunden der Ptolemäerzeit* (2 vols., Berlin and Leipzig, 1927-37). |
| *ZATW* | *Zeitschrift für die alttestamentliche Wissenschaft.* |
| *ZThK* | *Zeitschrift für Theologie und Kirche.* |

# BIBLIOGRAPHY

Barr, J., *The Semantics of Biblical Language* (Oxford, 1961).
— 'Hypostatization of Linguistic Phenomena in Modern Theological Interpretation', *JSS*, forthcoming.
Bauer, W., *A Greek-English Lexicon of the New Testament and other early Christian Literature* (translated and adapted from the German by W. F. Arndt and F. W. Gingrich, Cambridge, 1957).
Boman, T., *Hebrew Thought Compared with Greek* (London, 1960).
Burrows, M., 'Thy Kingdom Come', *JBL* lxxiv (1955) 1-8.
Caird, G. B., *The Apostolic Age* (London, 1955).
Cullmann, O., *Christus und die Zeit* (Zollikon, 1946); E.T., *Christ and Time* (London, 1951).
Delling, G., *Das Zeitverständnis des neuen Testaments* (Gütersloh, 1940).
— Article καιρός in *TWNT* iii, 456-65.
Dittenberger, W., *Sylloge inscriptionum graecarum* (3rd ed., 4 vols., Leipzig, 1915-24).
Dobschütz, E. von, 'Zeit und Raum im Denken des Urchristentums', *JBL* xli (1922) 212-23.
Eichrodt, W., 'Heilserfahrung und Zeitverständnis im alten Testament', *ThZ* xii (1956) 103-125.
Jackson, F. J. Foakes, and Lake, Kirsopp, *The Beginnings of Christianity: Part 1, The Acts of the Apostles* (5 vols., London, 1920-33).
M. Jastrow, *A Dictionary of the Targumim, the Talmud Babli and Yerushalmi, and the Midrashic literature* (London, 1903).
Jenni, E., 'Das Wort ʿolam im alten Testament', *ZATW* lxiv (1952) 197-248, lxv (1953) 1-35.
Knight, G. A. F., *A Christian Theology of the Old Testament* (London, 1959).
Lackeit, C., *Aion. Zeit und Ewigkeit in Sprache und Religion der Griechen.* 1ter. Teil: *Sprache* (diss. Königsberg, 1916).
McIntyre, J., *The Christian Doctrine of History* (Edinburgh, 1957).
Marsh, J., *The Fulness of Time* (London, 1952).
— 'Time' in A. Richardson (ed.), *A Theological Word Book of the Bible* (London, 1950), pp. 258-267.
Minear, P. S., *Eyes of Faith* (Philadelphia, 1946).
Murtonen, A., 'The Chronology of the Old Testament', *StTheol* viii (1954) 133-7.
Orelli, C. von, *Die hebräischen Synonyma der Zeit und Ewigkeit* (Leipzig, 1871).
Pfister, F., 'Kairos und Symmetrie', *Festgabe für H. Bulle, Würzburger Studien zur Altertumswissenschaft*, Hft. 13 (Stuttgart, 1938), pp. 131-50.
Pidoux, G., 'A propos de la notion biblique du temps', *RThPh* ii (1952) 120-5.
Preisigke, F., *Wörterbuch der griechischen Papyrusurkunden* (2 vols., Berlin, 1925-7).
Rad, G. von, *Theologie des alten Testaments* (2 vols., Munich, 1957 and 1960).

Ratschow, C. H., 'Anmerkungen zur theologischen Auffassung des Zeit-problems', *ZThK* li (1954) 360-87.

Robinson, H. Wheeler, *Inspiration and Revelation in the Old Testament* (Oxford, 1946).

Robinson, J. A. T., *In the End, God . . .* (London, 1950).

— *The Body. A Study in Pauline Theology* (London, 1952).

Sasse, H., Article αἰών in *TWNT* i. 197-209.

# INDEX OF BIBLICAL REFERENCES

*(in the OT, numeration normally follows the Hebrew text)*